CHILTON'S Guide
to
Large Appliance
Repair and Maintenance

CHILTON'S Guide
to
Large Appliance
Repair and Maintenance

Gene B. Williams

Chilton Book Company

Radnor, Pennsylvania

Copyright © 1986 by Gene B. Williams
All Rights Reserved
Published in Radnor, Pennsylvania 19089, by Chilton Book Company

Photos by Gene B. Williams and Deke Barker
Illustrations by John Kwiechen, Vicki Orza, Prue Carrico,
and Adrian J. Ornik, except where source is otherwise noted.

Manufactured in the United States of America

Library of Congress Cataloging-in-Publication Data

Williams, Gene B.
     Chilton's guide to large appliance repair and maintenance.
     Includes index.
     1. Household appliances, Electric—Maintenance and repair—Amateurs' manuals.
I. Chilton Book Company.   II. Title.
III. Title: Guide to large appliance repair and maintenance.
TK9901.W474   1986   683'.88   86-47608
ISBN 0-8019-7687-1 (pbk.)

SAFETY NOTICE
Proper service and repair procedures are vital to safe, reliable operation of all appliances, as well as to the personal safety of those performing repairs and using the appliances. This book gives instructions for safe, effective methods, and the warnings must be heeded. Follow standard safety practices at all times to eliminate the possibility of personal injury or damage to the appliance.

Techniques, tools, and parts—as well as the skill and experience of the individual performing the work—vary widely. It is not possible to anticipate all the conceivable ways or conditions under which appliances may be serviced, or to provide cautions as to all possible hazards. Standard and accepted safety precautions and equipment, along with all local electrical safety codes, must be observed. Heed all warnings when handling toxic or flammable fluids; wear safety goggles when cutting, grinding, chiseling, prying, or performing any other process that can cause material removal or projectiles. Some procedures require the use of tools specially designed for a specific purpose: do not substitute another tool for one designated.

ASBESTOS WARNING: Some older appliances may contain asbestos insulation. Asbestos is a known carcinogen; if you encounter it, do not handle it or breathe its dust. If an asbestos lining is flaking or deteriorating, dispose of the appliance promptly.

3 4 5 6 7 8 9 0   5 4 3 2 1 0 9 8 7

Contents

CHILTON'S Guide
to
Large Appliance
Repair and Maintenance

Introduction

Home appliances are usually touted as "labor saving devices." Salespeople and advertisements often tell consumers how much time and energy they can save each day by using these handy helpers. The pitch is that household appliances simplify chores. For once, the advertisements are not overstating things. Yesterday's luxury item is considered a necessity in modern life.

When was the last time you cleaned your clothes with a tub and scrub board, with water that had been heated over an open fire? When did you last try to bake a cake on a wood stove, or preserve your food in snow?

Hundreds of household appliances help us each day to perform chores more easily, quickly, and usually better. Sure, you could manage to live quite well without them, but would you want to? It would be like living on a permanent camping trip!

Our clothes are cleaned with appliances. Our food is kept fresh with appliances. The air in your home is kept comfortably warm or cool with appliances. Almost every part of our lives is touched in one respect or another with appliances.

Some of these devices are designed to perform chores while you sleep, or in your absence. As the clothes wash and spin dry, you can go into the other room to attend to other things; you can go grocery shopping while the dishwasher cleans the dinner dishes.

Labor-saving devices once considered to be tools exclusively for use in industry are now commonplace in private homes, all the way from electric pencil sharpeners and up to computerized washers and air conditioning units.

The home workshop can readily be supplied with numerous smaller "appliances," such as electric drills, sanders, saws, air compressors, and test meters. These tools can also be used to fix any of your *other* appliances whenever *they* malfunction. (These are covered in more detail in the companion book, *Chilton's Guide to Small Appliance Repair and Maintenance.*

We live in what has been called a "throw away" society. And that doesn't just mean that we throw away a lot of paper and other trash. We routinely toss out a variety of things because they're not worth the bother to keep or fix. Although this attitude is more common with small appliances, it's not all that uncommon to see someone ready to toss out a washer, dryer, or refrigerator because it has ceased to function properly, and the owner thinks that it isn't worth the hefty service charge to put it back into shape.

Not long ago a friend of mine was about to spend (waste, actually) over $300 to replace a malfunctioning dryer. It no longer heated and it squeaked when the drum was spinning. Since the machine was over ten years old, he just assumed that it wasn't worth the cost of fixing.

The problem turned out to be a faulty heating element (about $7 and 3 minutes to replace), and a drive belt that had slipped (free and less than 1 minute).

In another case, a $900 refrigerator was given away because it leaked water onto the floor, had a build-up of ice inside, and seemed to be running constantly. The first two problems were cured within a few seconds after a piece of food was removed from the drip tube. The second took more work—about $20 and a half hour to replace the seal around the door.

Household small appliances generally are priced at $100 or less. Here it is often tempting to buy a shiny new replacement rather than take the effort to make the diagnosis and repair. Professional repair shops today normally have a base or minimum charge of $30 to $50, just to open up a malfunctioning device to attempt to see what is wrong. So if a $24 electric coffee maker acts up, and the service charge to fix it will be $35, it is cheaper to buy a new one than to hire someone to fix up the old. Out goes the old coffee maker, and you plunk down $24 or more for a new one.

With a large appliance that costs hundreds of dollars, the decision isn't quite as easy to make. At this point many people try to come up with a market value of the appliance and compare this to the cost of repair.

The fact is that the whole problem with the appliance might be a $7 heating element, or a 50¢ switch. Worse yet, the problem might be nothing more than a belt that has slipped off its track, or a wire that has come loose that simply needs to be pushed back into place. Total cost to repair the unit in this case is nothing but a few minutes of your time.

Most of the cost of repairing an appliance is in the labor. It could easily cost you $100 or more to fix that dryer, even though the actual cost of replacement parts is only a couple of dollars. If the repair shop has a minimum charge (quite common), the actual cost of repairing the unit will be zero, with the entire charge being wrapped up in the few minutes of labor needed to take out a couple of screws, unplug the old heater element, plug in the new one, and tighten the screws again. (Or to tip the dryer a little and give the belt a push.)

The purpose of this book is to save you those labor charges. Why pay someone $40 an hour to take out a couple of screws? Especially when repairing most problems is so very easy.

Many people hesitate to attempt repairs on home appliances. The concept of do-it-yourself repair is often bothersome, especially to people who feel they are not mechanically inclined. This apprehension might be justified when it comes to repairing an automobile or a leaking roof. If the work on those projects is done improperly, it is extremely costly to replace a car or home if damaged beyond salvage by incompetent workmanship.

Going back to the clothes dryer example—assume that the device no longer heats. Chances are good that the only problem is a faulty heating element. Cost of a replacement is just a few dollars. To make the replacement, you usually remove four screws from the back of the dryer, unplug and slide out the old element, plug in the new, and put the screws back in. Total time for the job: about 3 minutes. The savings you'll realize: about $75.

Get out your calculator and figure it out. At $75 for a 3-minute job, that works out to be $1500 per hour! Even using more realistic figures—spending a few minutes of your time and $7 on parts to save yourself a $75 service call or the $300 or so for a new dryer—it still profits you nicely.

The important thing to keep in mind is that for the majority of large appliance malfunctions, you *can* make a successful repair, even if you don't have a mechanical background. It's not as difficult as you might think. Diagnosis of a problem is usually easy. When it's not, a few simple tests will often make it obvious. And you don't need a bench full of fancy equipment and tools to get the job done. Nor do you need years of special training. Try it once and you'll realize just how easy it is. The more you do, the easier it gets—and the more you save. (You might even begin to wonder why in the world we've become a "throw-away society".)

There will be times, however, when your attempts at repair will be useless. You might find that the repair is either impossible, or requires more time, effort and actual cost (for parts) than the price of a new device. Even then you don't lose out.

First, you can use these ailing appliances as a sort of training step for learning more about do-it-yourself projects. The next time something needs fixing, you'll be better prepared to handle it.

Second, and almost as important, in taking apart the malfunctioning device you will have provided yourself with a "spare parts" supply from the old unit. Those can be of obvious value if you buy a new one of the same make and model. What many people don't realize is that quite often the salvaged parts can be used in other appliances. (Electric cords are an excellent example of this universal adaptability.)

Beyond just saving money, there are other reasons for attempting to fix a broken appliance yourself. The item may have been a gift that you treasure. It may not be worth the price of professional repair, but if at all possible you'd still like to keep it around and operational.

In some instances, a favored appliance may be an old model that suits you just fine. You prefer the way the old one did the job and the "quality of workmanship of days gone by." You'd much rather have the familiar appliance back in good working order than have to make do with a newer or more complex version you dislike.

In almost all cases, appliance repair is a case of nothing ventured, nothing gained. And you are in a strictly no loss situation, since the loss is already there if the appliance can't be whipped back into shape through your own efforts. The most it can cost you is some minutes of your time to find out if you can successfully make the repair.

With a few simple hand tools, common sense, and this book, you'll be able to diagnose and fix most of the common problems that can interrupt the smooth and reliable operation of these everyday conveniences. You'll also learn how to carry out some simple maintenance steps to reduce the number of times that repair becomes necessary.

In short, the cost of this book could be one of the more valuable investments you've ever made! You can save its cost many times over with your first job.

In Section One you will be given all the basics you need to successfully repair and maintain the major appliances in and around your home. Section Two is a guide to troubleshooting and repair of various appliances. The eight most common large appliances are presented: refrigerators and freezers, washers and dryers, ranges, dishwashers, air conditioners, and heaters.

Chapter 1 describes the basic tools needed for repair jobs and how to use them. It also lists some of the supplies you might wish to keep on hand if you intend to become a serious do-it-yourselfer.

Chapter 2 is dedicated to the subject of safety. *Nothing* is more important than safety. The chapter is fairly short and to the point, so read it, and then read it again. It's the single most important chapter in the book. Every year people lose their lives because they consider safe procedures to be too boring to learn. Repeat: there is *nothing* more important than safety.

Since almost all appliances are electrically operated, Chapter 3 is devoted to a description of home electrical codes and practices and to demonstrations of how best to handle electrical repairs. It is filled with practical information to help you avoid making costly and possibly dangerous mistakes. Put this information to use and you are highly unlikely to run into trouble while working on any electrical, motor-driven, or mechanical device.

Chapter 4 describes the diagnosis and repair principles for all large appliances. These fundamental procedures will let you handle all but the most complex repairs on any device. This chapter will be of help even if your appliance is not individually listed in Section Two.

Electric motors are covered completely in Chapter 5. These are the power sources for most appliances. You'll discover that even though many of the electric motors used are sealed, some can be opened and brought back to life again rather easily. (You'll also find that many times a motor problem is nothing more than a loose connector.) Inexpensive sources for replacement motors are listed, in the event you are unable to get the original one working again.

Chapter 6 covers all the heating elements used in appliances such as dryers, or any other device where providing heat is one of its functions. You will be shown how to isolate the malfunction to a particular part, how to obtain replacements, and then how to install the replacement properly.

Chapter 7 tells how to maintain appliances to help reduce the number of times that repairs are needed. Sound maintenance practice can go a long way toward keeping your convenience item operating properly.

In Section Two, you will be guided step-by-step to troubleshooting and repair of eight major appliances. For each appliance there is a discussion of the basic design and operation, a review of common breakdowns, and troubleshooting tips to assist in problem identification and diagnosis. Specific information is then provided for replacement and/or repair of the defect.

If the appliance that is giving you trouble isn't specifically listed, refer first to Section One. Chances are, you'll find exactly what you need to know here. You can also look up a similar appliance in Section Two for more specific information. (If it's a small appliance that is giving you trouble, be sure to see the companion book, *Chilton's Guide to Small Appliance Repair and Maintenance.*

Before anything else, your first step should be to read through this book completely. Section One is of special importance since it provides all the basic information you'll need for the job at hand, and also gives you the safety tips that will protect both you and the device on which you are working.

Many people will be tempted to skip Section One. Don't.

Even more people will ignore Section Two except for those parts that are of immediate relevance. That, too, is a mistake. Even if you're working on a clothes dryer, the hints on diagnosis and repair of an electric furnace can be handy.

After reading the book, study Chapter 2 again. The importance of safety can hardly be stressed enough. If you can't understand how important safety is and can't spare the time to study Chapter 2 before beginning, toss out the malfunctioning appliance and buy a new one. You'll save money (because you can't be bothered to learn proper repair procedures), and possibly your life.

Steps 3, 4 and 5 on the list are actually more like three parts of a single step. Carry out the diagnostic steps and read the appropriate part in Section Two with its specific diagnostics, tests, and hints for that particular appliance. Once this has been done you'll know what is causing the

---

The First Steps

1. *Read the book completely and thoroughly, particularly Section One.*

2. *Study Chapter 2 once again (for safety tips) before you begin. Read Chapter 3 if applicable.*

3. *Perform the basic diagnostics steps of Chapter 4.*

4. *Read the applicable repair description in Section Two.*

5. *Complete the tests on your unit to find the defective part.*

6. *Repair or replace that bad part.*

7. *Use your appliance!*

---

trouble and you can proceed with the repair or replacement of the defective part.

This brings you to the nicest part of it. You can put away your checkbook and stop watching the ads for appliance sales. The one you were about to toss in the trash will be functioning again.

Section One

Before You Begin

Chapter 1
The Tools and Workshop

A few hundred years ago, the average person was capable of making the needed repairs to the simple machinery used around the home or farm. Despite how complex machinery has become, the same holds true today. The only real difference is that many people don't realize that *they can do it*.

As more sophisticated equipment moves into our private homes, it's easy to think that the task of repair requires more knowledge and experience than the average owner has. As much as that owner wants to avoid the high cost of repair, he or she just doesn't feel capable of handling the job. It seems too complicated for the homeowner who rarely handles a screwdriver let alone the insides of a washing machine. The owner with this common attitude may spend hundreds of dollars to have a professional come out to make the repairs, often when the owner can easily do them. (You'll find that the majority of repairs are very simple. Have you ever known someone who paid $50 to have a repair technician come out just to push in the plug? It happens all the time.)

Repairs to appliances are not as difficult as you might think. Once you've overcome that first major hurdle of realizing that you *can* get the job done as well as—and often better than—that professional, you've already accomplished at least 75% of the job.

The goal of this book is to give you the remaining 25%. It will show you some of the steps that professional technicians use to test appliances, and to quickly determine what part has malfunctioned.

GENERAL PURPOSE TOOLS

Before you attempt any repair work, it is necessary to acquire a few basic hand tools. You probably already have most of these in your toolbox. If you don't, it's past time that you made the investment. These things are too useful to worry about the few dollars they will cost. It's a one-time investment if you buy wisely. It's also an investment that will pay off many times over.

As a simple example, imagine that the outlet for your dryer has become faulty. You could pay a professional $50 or more (plus parts, which are charged to you after a hefty markup has been applied) to replace it. Or you could dig out that $2 screwdriver and do it yourself. That's a savings of $48—and you get to keep the screwdriver.

Appliances are held together with various types of screws, nuts, and bolts, along with other fasteners. Once inside, you'll also find that many of the components are held in place in much the same way. Before you can do any servicing, you'll have to acquire the tools needed to get inside and to work with the components.

You will need a few standard screwdrivers with tips of various sizes. All must have insulated handles, since most appliances are electrically operated. Different sizes of screwdrivers are required because appliances use different sizes of screws. Using the wrong size screwdriver can cause several problems. First, it may not work at all. Second, it may damage the screw, making it extremely difficult or impossible to re-

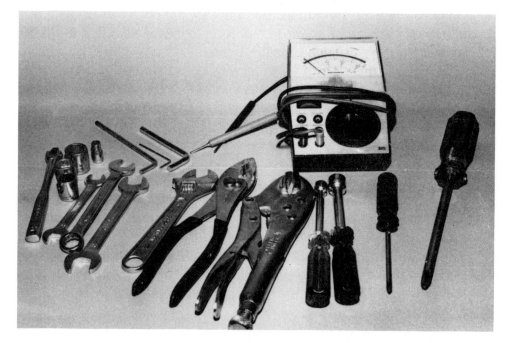

FIG. 1-1 Basic tools needed. From left, socket set, wrenches, allen wrenches, crescent wrench, vise grip, nut drivers, screwdrivers, and VOM.

Basic Tools

Screwdrivers
    (blade and Phillips)

Wrenches

Socket set

Allen wrenches

Nut drivers

Pliers

Needlenose pliers

Wire cutters

Knife

Soldering tools

Multimeter
    (volt-ohm-milliammeter)

move even if you get smart and switch to the proper size screwdriver for the job. Third, you can damage the appliance. Fourth, and most important, you can injure yourself.

Use the right screwdriver for the job. The few extra dollars you'll spend for a complete screwdriver set will pay off.

Another type of screw head is called a Phillips, which is like a four-sided star. You'll need at least one with a fairly small head. It's also handy to have a Phillips screwdriver with a larger head on it. A few large appliances require a Phillips screwdriver with a very small head. Screws of this type are sometimes used to hold smaller components in place. Be sure that the screwdriver has an insulated handle. And be sure that you use the right size screwdriver for the screw.

Somewhat like a Phillips head is a relatively new design called a Torx. This kind of screw head is highly resistant to stripping. Almost certainly more and more manufacturers will begin to use this kind of screw. If your appliance uses this kind of screwhead, *do not* try to remove it with anything other than a Torx driver. Using a Phillips or blade screwdriver will damage the screw, the driver, and possibly the appliance.

Occasionally you'll come across other types of screw or bolt heads. The most common of these have a hex head, either indented (use an allen wrench) or with a nut-like head (use a wrench or nutdriver). In many

BLADE   PHILLIPS   HEX   TORX

FIG. 1-2 Screwhead types.

of these cases, the screw or bolt will have a head that gives you a choice of tools. Commonly, the head will have a slot for a screwdriver blade, surrounded by a hexagonal nut that can be turned with a wrench, socket, or nutdriver. When given this choice, it's always preferable to use the nut head instead of the slot. The job is easier and safer this way.

A set of wrenches will often be needed to get inside the appliance, and sometimes a wrench will be the only way to take care of certain jobs. For example, replacing a belt often requires loosening and tightening the motor mount bolts. If you don't tighten the bolt completely, the belt will soon be flopping again.

An alternative to wrenches (but not necessarily a replacement) is a socket set. These days you can find inexpensive sets in many places. A complete 40-piece set might cost as little as $5. These tools are obviously of lower quality, but are certainly sufficient for the occasional handyman.

Nut drivers have been mentioned several times. If you intend to do much repair work in your shop, another useful acquisition is a set of nut drivers. These are like fixed socket wrenches, or like a combination of a screwdriver and a socket.

A pair of pliers with plastic insulated handles and a pair of needle-nosed pliers will be of help to you. These are used for gripping. *Do not* use them in place of a proper wrench or socket.

As always, it's best to use the tool meant for the job. A separate wire clippers is generally better than the one built into a pliers. Although you can use a sharp knife to strip away the insulation from a wire, a wire strippers will do the job better, quicker, and more safely. (Even so, a sharp pocket knife is something no tool kit should be without.)

Another convenient combination of tools for the home workshop is the small soldering iron and a desoldering tool (often called a "solder sucker").

FIG. 1-3 This kind of fastener gives you a choice of tools. Whenever possible, use a nut-driver instead of a screwdriver.

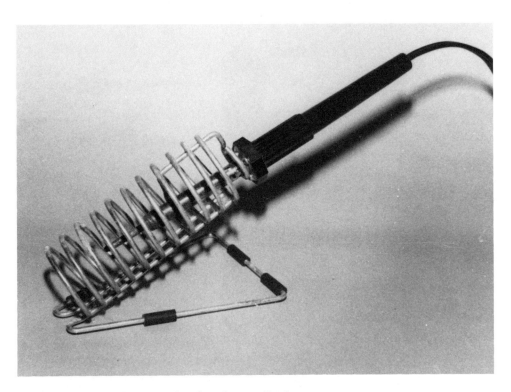

FIG. 1-4 This soldering tool is fine for small jobs.

For some appliance work, a heavy soldering iron or gun of 100 watts is best. This type of iron is too hot for working on sensitive solid-state electronic circuits such as the transistors and IC chips in a television set, but the extra heat is often needed for effective soldering work on many home appliances.

When buying a soldering iron or solder, tell the person at the counter what type of work you will be using it for. Most shop personnel are familiar with the correct style and model you should be using.

The solder sucker, which literally sucks up excess solder from a heated joint, is used to remove old solder when you are attempting to replace a part or disconnect a wire so that meter readings may be taken properly. These tools come in several different types, from a simple squeeze bulb to a fancy spring-loaded heavy-duty model. If you have to disconnect a soldered part, first you must melt the old solder joint and, while the joint is still hot, use the desoldering tool to remove the melted solder. The component or wire can then be removed from its original position easily.

Before attempting to use a soldering iron, study carefully the pages on its proper use in Chapter 3. Better yet, read this section, then practice, practice, practice! Soldering is fairly easy, but it must be done correctly.

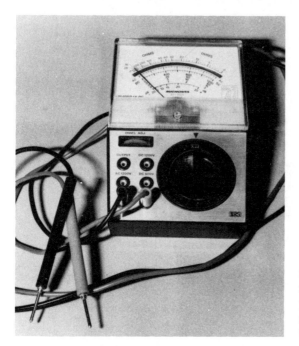

FIG. 1-5 The VOM is one of the most important tools for the do-it-yourselfer. You don't have to spend a lot of money to get one that works.

In electrical work, always use resin core solder. Acid core solder, used in some forms of construction requiring the binding of metals, is corrosive. If used for making electrical connections, the joint will eventually deteriorate and the electrical contact will be destroyed.

One tool you can hardly afford to be without—a VOM (volt-ohm-milliammeter)—is discussed at the end of this chapter. For the moment, jot this down as one of the "must get" tools, and if you aren't already familiar with its operation (and even if you are), turn to the last section in this chapter for more information.

As you become more sophisticated in repair and maintenance work, you may wish to obtain other tools, such as a small electric drill, and possibly a set of clamps, vises, and holders in order to work on jobs requiring a "third hand." A third hand is a tool made up of alligator clips and arms built into a stand; it aids in holding wires and other small objects that need soldering or other attention.

Heating elements in particular can't be soldered. While in operation they get too hot, and regular solder would simply melt off. Consequently, you may need a good crimping tool, or possibly a small torch and silver solder. More information on how these are used is contained in Chapter 5, "Heating Elements."

SUPPLIES

You don't need a huge stock of supplies to carry out successful repairs. As mentioned in the Introduction, you can slowly build up your collection of spare parts as you go along by scavenging unfixable appliances. Meanwhile, the local parts supply store will serve very well as an out-of-home stockroom.

Your only real need is for supplies that will be used for many different repairs. Several of these are described below.

A roll of good quality electrical tape can be indispensible around the home. This is the plastic-coated insulating tape, not what is often called "friction tape." If you don't already make a habit of having plenty of electrical tape around the home, once you get some you'll wonder how you got along without it.

For cleaning and preventive maintenance of appliances, the do-it-yourselfer should have on hand a bottle of denatured alcohol (not rubbing alcohol) and cotton pads and swabs. The more pure the alcohol, the better. You can find it on the shelves in most drugstores with a purity up to about 90%. This is fine for most jobs. Even so, you may wish to go to the trouble of getting a better grade, such as 99%. The difference in cost is very small and assures that you won't be contaminating when your goal is cleaning.

A can of spray lubricant, such as WD-40, and a supply of light machine oil can help in preventing costly appliance breakdowns if used properly. Used improperly both can cause more problems than they'll cure. The trick is to use as little lubricant as possible, and to thoroughly clean away any excess afterwards.

Also desirable is a supply of various connectors.

You should be able to find connector sets at most hardware and electronic stores. These contain several each of different kinds of con-

---

Handy Supplies

Electrical tape

Denatured alcohol, cotton pads and swabs

Spray lubricant

Connectors of various types

Wire, 10 to 16 gauge

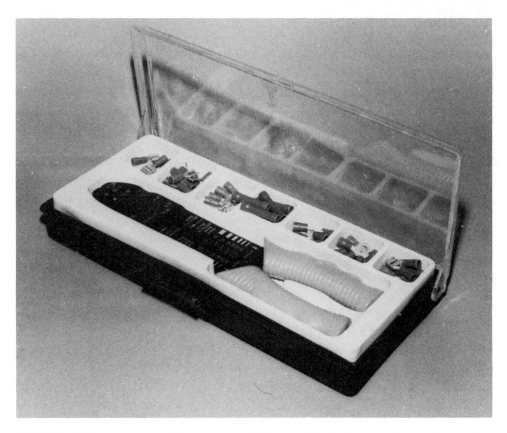

FIG. 1-6 This kit comes with a variety of different kinds of connectors and a multi-purpose tool.

nectors and are a good way to get started. Later on you can get packages of specific connectors. (It's almost always worth getting a package rather than single connectors, unless you need a specific one.)

For about $10 you can buy a more complete set with connectors and a wire stripper/crimper/cutter tool. For occasional jobs, these combination tools work just fine. If you're planning to do a number of repairs, it's worth it to pay a little extra for a higher-quality set.

A supply of different sizes of wire is also useful. Not just any wire will do. You can often (but not always) replace a smaller gauge wire with a wire of larger gauge. The reverse isn't true.

The wire that carries a tiny control voltage might be just a few strands of thin copper covered by a light coating of insulation. The wires going to the heating elements of an electric range are probably heavy wire with matching insulation. If you try to replace the heater element wire with the small one for a control, the best that will happen is that the wire will melt and everything will stop. Hopefully nobody would be silly enough to try such a thing. However, it's all too common for someone to replace a heavy 10-gauge wire with 14-gauge wire because that happens to be at hand. Failure in this case won't be as quick, but that makes it all the more dangerous. That too small wire will be getting very hot, possibly hot enough to melt away the protective insulation. And all of a sudden you have a deadly 240 volts just waiting for you to touch the wire. The voltage may even be conducted through the metal body of the appliance.

A well-equipped work bench will have small spools of wire from gauges 10 to 16. If buying a spool each of the different wire sizes is more than you care to spend, don't worry. Many hardware stores carry large spools of wire and allow the customers to buy whatever length is needed. It is simply more convenient to have a supply at home.

The best way is to get things as you need them, but in slightly larger quantities than are required for the job at hand. For example, if you have a need for 16 inches of 10-gauge wire, get an extra few of feet at the same time. The additional wire will cost you very little. It allows you to make mistakes, and if you don't make mistakes, it gives you some extra wire for the next time that size wire is needed. You might also find that same gauge wire in a small spool of 25 or 50 feet for about the same cost as the few feet you actually need.

THE VOLT-OHM METER (VOM)

To test for electrical voltages, measure component values, and check for continuity (continuous wiring contact from one point to another), you will need a multimeter (volt-ohm meter). Testing a wall outlet *can* be done easily with a tester made specially for this purpose, and some technicians prefer this simpler tester. Others feel that the only viable way to test *any* voltage is with a meter.

Fig. 1-5 is a photograph of a simple volt-ohm-milliammeter, or VOM. Today, some pretty fancy digital VOMs are available from hardware and electronic supply stores for less than $50. Simple models with a needle and scales are available at prices less than $15.

Although a more expensive meter is usually better, you can get by very well with one of the very simple meters. Almost all meters made today, including those $10 specials, are accurate enough for your purposes.

Whichever VOM you get, it should be able to measure at least voltage and resistance. Unless the meter has built-in automatic switching, the meter should have several ranges for each kind of reading.

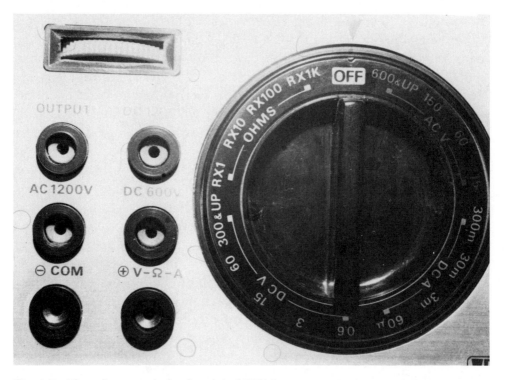

Fig. 1-7 The selector switch of a typical VOM.

The voltages you will be measuring will be somewhere between 5 to 50 volt dc (vdc), and 120 or 240 volt ac (vac). The meter you choose doesn't necessarily have to provide exactly those settings. For example, I have one meter that offers dc voltage choices of 0.6, 3, 15, 60 and 300 + for a setting. There is no specific setting for 12 vdc, but this doesn't matter. By putting the meter in the 15 vdc range, 12 vdc can be measured very accurately.

The meter you purchase should also be capable of measuring resistance (in ohms — the "O" of VOM). To many newcomers, reading ohms is useful only to check the value of resistors. This is actually the least useful way to use a VOM. (Resistors are already color-coded as to value anyway.) Resistance tests are useful for checking such things as continuity (a complete and continuous electronic path).

A third function on many VOMs is current reading. An analogy often used is one that compares electricity in a wire or circuit to water in a hose. In this analogy, current would be like the number of gallons per minute of water, with voltage being the force or pressure of the water. The milliammeter function of a VOM measures the quantity of electricity in thousandths of an amp. It's very rare to need this function when testing or repairing appliances. Usually, if you need to measure current flow, that flow will be greater than 1 amp and could be as high as 50 amps. A special meter is needed for this.

Carefully read the instruction manual that came with your VOM. The operation of each model and style of meter is slightly different. Also, meter scales read differently, so be sure to study your meter's scale or digital readouts so that you can interpret the meter or dial readings the instrument gives.

A VOM can be damaged internally if it is set to read resistance (the ohms scale) and then the probes are connected to a voltage source. Be sure you know what you want to measure (volts, ohms, current) *before* connecting the probes or turning on any power source. It can also be damaged if you use the correct setting by the incorrect range (set to read 3 vdc while probing a 60 vdc circuit).

The following sections will give you the basics of using a VOM. If you are not familiar with its use, be sure to practice with the meter and learn its functions well before probing inside one of your appliances. Take readings of the various outlets in your home to check the ac voltage scales. Test old batteries to get used to the dc scales. Use unplugged extension cords to get practice in testing for continuity — to see if the wire is broken (or "open").

In short, get used to the meter before you have a real need to use it.

TESTING FOR VOLTAGE

As you'll learn in this book, finding the cause of a malfunction (diagnostics) is nothing more than a process of elimination. There are only so many reasons why something fails to work as it should.

Assume that an appliance has failed completely. There are just four places that the problem could be. It's possible that there is no power coming into your house; or that the trouble is between the house service box and the outlet; or it might be getting that far but can't get into the device due to a faulty power cord; or it could even be something in the appliance itself.

The VOM will help you to track it down quickly.

Before you even get out the meter, eliminate the obvious, such as the appliance being unplugged. If it *is* plugged in (are you *sure?*) the VOM can be used to test for voltage at the outlet.

It's possible to test the outlet by plugging in a lamp or another appliance. However, there are times when the incoming voltage is present, but not high enough in value to operate that particular appliance. A lamp, for example, might look normal, when in fact the incoming volt-

FIG. 1-8 Testing a wall outlet for power.

age is insufficient for your device. Testing with a VOM will tell you for sure. If your area is prone to "brown-outs," using a meter to test the outlet is especially important.

To test a wall outlet, set the meter to the correct ac voltage range (120 vac for a standard wall outlet, and 240 vac for heavier appliances). Keep in mind that you are probing potentially deadly voltage. Hold the probes by the insulated handles only! Do not touch the metal of the probes.

Very carefully insert the two probes into the flat slots of the outlet. The meter should read 117 volts on the scale, or very close to it. With a grounded outlet (two flat slots and a round one), you should be able to get the same reading between the ground (the round hole) and the slot with the incoming line, but not between the ground and the other slot. (If you get a reading between the ground hole and both flat slots, it's time to call in an electrician.)

With a 240-volt outlet, set the meter to read in the correct range and probe the slots in the same way. If everything is normal, you'll get a reading of 117 volts between the lower slot and each of the two upper slots, and a reading of about 234 volts between the two upper slots.

If you get the proper readings at the outlet, you've just eliminated

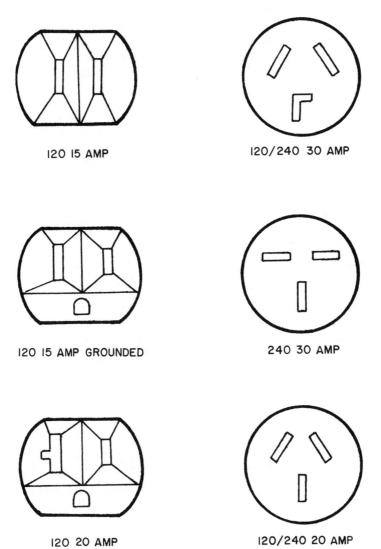

120 15 AMP

120/240 30 AMP

120 15 AMP GROUNDED

240 30 AMP

120 20 AMP

120/240 20 AMP

FIG. 1-9 Typical outlets; standard 120 vac wall outlets and 240 vac outlets.

two of the four possibilities. In fact, even if you *don't* get the proper readings—assuming you've used the meter correctly—two of the four have been eliminated. If the readings are correct, then power is getting to the outlet and the problem is either in the cord or in the appliance. If the readings aren't correct, then the appliance is probably fine (or it has a short circuit that is causing the fuse or circuit breaker to blow). The problem is either with the power company, or someplace between the service box and the outlet.

In a few cases, an appliance will have a built-in power supply that takes the incoming 120 vac and converts it to other values of ac, or to various values of dc. Quite often these are marked on the appliance. Measuring is the same as measuring voltage across any wires. Keep in mind that electricity has to have a complete path in order to work. It has to come in somewhere, go through the circuit, and then exit in a complete path back to the supply.

When measuring dc voltage or current, the polarity of the test leads is important. Direct current (dc) is current moving from point A to point B in a circuit in a single direction. Theory assumes that current moves from source (the "hot" side of a circuit) to ground. The common or ground lead of the meter goes to the ground side of the circuit, the red probe goes to the hot side to read dc voltage or dc current. Check, and then recheck, the polarity of the test leads connecting voltage to a circuit before applying power. If hooked up backwards, a meter with a moving needle will have the needle deflect downward, in a reverse direction. Digital meters will display a − (negative) value. To get correct readings, just reverse the polarity of the probes.

Check (and then recheck) the *range* of your meter before applying power. You can damage the VOM by trying to read voltage from a supposed 50-vdc source when your meter selector switch is set to the dc 0-5 volt scale.

It is not necessary to observe polarity when measuring ac voltages. Alternating current moves back and forth in a wire or circuit, first in one direction and then the other. Like resistance, ac voltage and current can be checked without worrying about which probe is on the negative and which is on the positive terminal of the power source.

When testing ac voltage, be sure that you locate and identify the specific ac connections before probing. For dc, it's generally not as critical. You can often safely touch the black probe to any known ground and then use the red probe to take the readings. *Never* touch the black probe to the chassis (and the red to the suspected spot) to test for ac voltage. The meter becomes a part of a complete circuit in this case and can make the entire appliance deadly to touch. In an appliance where both ac and dc are present, be careful that you know which wires and connectors carry which voltage, and at what value, before probing.

TESTING FOR CONTINUITY AND RESISTANCE

If there is power at the outlet but the appliance is dead, the trouble could be a damaged power cord. It is sometimes possible to probe the connectors where the ac voltage enters the appliance. Even so, it's best to test the cord and connectors by checking for continuity.

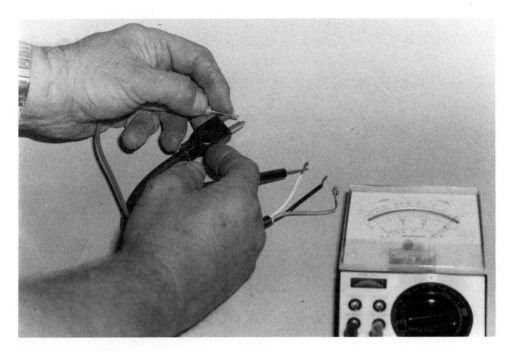

FIG. 1-10 Testing a power cord for continuity.

As mentioned in the section above, for electricity to work, it must have a complete and continuous path. If that path is broken anywhere, current stops flowing. In a power cord, if one of the two wires is broken or has come loose from the connector, the pathway is no longer complete, and there is no continuity.

The resistance (ohms) setting of your VOM will let you test for continuity. In simple terms, if the wire being tested is intact, you'll get a reading very near zero ohms (no resistance to current flow). For a meter with a needle, this will show by the needle swinging full-scale all the way across the meter. If there is a break in the wire, the needle won't move, indicating that the resistance is infinite (no continuous path, and no continuity).

Testing for continuity is one of the best diagnostic tests you can make in appliance repair. First be sure that the appliance being tested is turned off and unplugged. If you're testing the power cord, you'll have to be able to access both sides of the cord. This means that if the cord doesn't unplug from the appliance itself, you'll have to open the appliance to get at the connectors inside.

Put the selector switch of your VOM into the resistance (ohms) range. The actual range isn't critical, but it is generally best to use the lowest setting possible. (This is usually the X1 range.) Touch the ends of the two meter leads together. The meter should indicate zero ohms, or no resistance. The meter is showing that there is an unbroken, continuous path from one probe to the other. If there is a slight reading, use the "zero adjust" control on the meter to set for zero.

Now touch one of the probes to one side of the wire and the other probe to the other side of the same wire. You should get a reading of zero ohms.

To test for a short circuit along the cord, touch one probe to one of the two wires, and the other probe to the other wire. A reading of infinity—which is what you want—shows you that there is no path between the two legs. Any reading at all shows that current can flow between the two wires—and it shouldn't.

In sequence, touch the other probe to each of the three prongs on the plug. *One* of the prongs should show continuity (zero ohms) to the wire at the other end, the others should show open (a reading of infinity, and no deflection on the meter).

This is also one swift way to trace a wire from one point to another, to see which wire is which in a circuit. When you read continuity, you know those two points are wired together someplace or other in the ap-

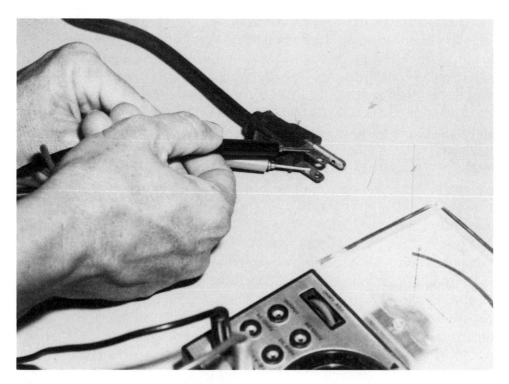

FIG. 1-11   Testing a power cord for a short circuit.

pliance. If your meter indicates "open" (no connection) from any of the three prongs to the other two, you have no short circuits in the cord.

In turn, move the probe at the open end of the cord to the other two wires, and make sure that each shows continuity (zero ohms) to *one* prong, and open (infinite resistance) to the other two. The same type of meter test can be made with one of the simpler push-on type of clamping replacement plugs, if the end of the cord away from the plug is also exposed. If it's a two wire plug, the VOM must indicate that you have continuity from one prong to one wire, and an open circuit from that same prong to the other wire. It must also indicate continuity from the second prong to the second wire, and an open from the second prong to the first wire.

Testing for resistance is an excellent way to find out if a particular component or element is starting to give out. A heater element, for example, can be tested easily with a meter. With the power off and the appliance unplugged, disconnect one side of the element. Set your VOM to read resistance and touch the probes to the two ends of the element. You should get a reading near zero ohms. If there is no meter deflection, the element has become "open," and there is a break in the wire. A reading that is not infinite (no deflection) but shows a high resistance is a sign that the element is starting to wear out. Either way, it's time to replace that element.

This same kind of testing can be used elsewhere in the appliance.

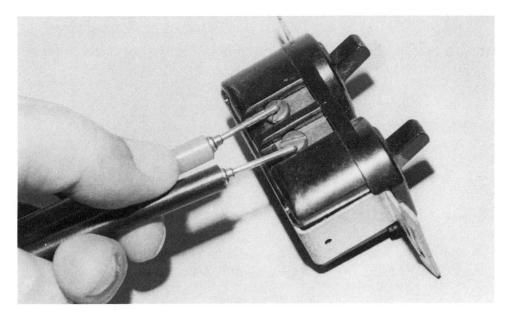

FIG. 1-12 Testing a switch with a VOM.

(Always be sure that the power is off and that the appliance is unplugged before you begin.) By touching the probes to two ends of a wire, you can determine if a complete path exists (continuity) or if there is a break in that circuit. It can also be used to test for short circuits by touching one probe to the chassis of the appliance and the other to the suspected component.

Even switches can be tested with the VOM. Find the connectors across the switch and touch these with the probes (again with the power off, the appliance unplugged, and the meter set to read ohms). With the switch in the off position, there should be no reading; and with it switched on, there should be a reading of zero ohms. If this doesn't happen, the switch is bad.

TESTING FOR CURRENT

Current is the amount of electricity in amps flowing through a circuit in order to make it work. To measure current the meter must become an actual part of the circuit. You can measure resistance or voltage by placing the meter across (in parallel) with the circuit, with one probe touching one end of a wire or on one terminal and the other probe touching the other end of the wire or a second terminal. Current can only be metered from within a circuit.

To give a practical demonstration, take a working circuit consisting of a battery-operated transistor radio. You can measure the voltage of the battery or of points in the circuit by putting the common (ground, or black) probe of your meter on the negative side of the battery, by moving the red (hot) probe to the positive terminal of the battery or to other ran-

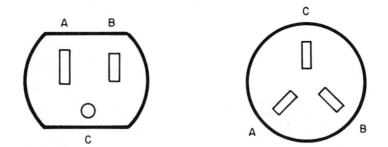

FIG. 1-13 On a 120-volt ac outlet, the reading between A and B and between B and C should be approximately 117 volts. The reading between A and C should be zero. For a 240-volt ac outlet, the reading between A and B should be approximately 234 volts, and the reading between A and C or B and C should be 117 volts.

dom circuits in the radio. Resistance may be read by disconnecting the battery (never read resistance while voltage is being applied), and connecting the two probes to any two different points in the radio circuit board.

In order to determine the amount of current used when the radio is operational, you must disconnect one terminal of the battery from the radio, while leaving the other battery terminal properly connected. Say you have disconnected the ground end of the battery, while leaving the positive terminal connected. Now be sure the meter selector is in its highest dc current position. Then connect the black (negative or common) probe to the negative terminal of the battery, and the red (hot) probe to the point in the radio where the negative battery terminal normally connects. The voltage from the battery will force a current to flow through the now completed circuit—from negative terminal of the battery through the on switch, then on through the radio, through the meter, and back to the hot or positive terminal of the battery. In this process, the current flow will cause the meter to read, and you will be able to determine the amount of current in amps, milliamps (thousandths of an amp) or microamps (millionths of an amp) that is being used to make the radio operate.

Most VOMs are capable of reading current only in the milliamp range. This is fine for measuring the amount of current consumed by a transistor radio. It's not enough to test for current in a "Fry Daddy." Another kind of meter, and an expensive one, is needed for that.

Fortunately, current measurements are seldom necessary in appliance repair. If you can quickly and accurately make resistance and voltage measurements, you will be competent to do most of the diagnostic work involved in basic small appliance troubleshooting.

Chapter 2
Safety

It's tempting to get out the tools and tear into that malfunctioning appliance. But even if you're an old hand at handling electricity and machines that use electricity, take a few moments to review the basics given in this chapter. (If you *are* an old hand, you'll also realize just how important this review is.)

If all appliances were operated by battery, there would be no real danger. The dc voltage from a battery isn't nearly as dangerous as is the ac voltage used by your appliances.

Some years ago the United States Navy responded to the potential dangers of ac voltage by conducting a very thorough study of the subject, particularly the effects of current on the human body. The results of the study showed that it takes surprisingly little current for ac voltage to cause harm. Just 1 milliamp (1/1000th of an amp) is enough to be felt. 10 milliamps (1/100th of an amp) is sufficient to cause muscle spasms and paralysis. At this point, the person being shocked will be unable to let go of the source of the shock, and is likely to actually grip it more tightly. If the current is increased to 100 milliamps (1/10th of an amp) and continues for more than just one second, the usual result is death. The most important muscle in your body, your heart, will become paralyzed. You can't live long under those circumstances.

This isn't meant to scare you — although it should. The purpose is merely to let you know that you are definitely vulnerable. A flow of electricity of just 1/10th of an amp can be fatal, and you're dealing with a much greater flow than just a tenth of an amp.

A standard wall outlet in your home will be protected by a circuit breaker or fuse that can handle 15 amps. Other outlets might be going through a 20 amp or larger breaker or fuse. Large appliances will normally go through a circuit breaker or fuse capable of handling between 30 and 100 amps, depending on the appliance.

Even that relatively small outlet, then, can easily carry 150 times the current needed to be fatal. And that is for an almost indefinite period of time. Before that breaker or fuse lets go, or the wires in the walls melt, the current flow for a few seconds can be almost limitless — thousands of times more than your body can tolerate.

There are other dangers as well, such as heat and the risk of injuring yourself physically through carelessness. As has been mentioned before, you *can* successfully handle most repairs yourself. You can also handle them in perfect safety if you just pay attention to a few basic rules.

The money you've saved in doing your own repairs will be of little benefit (except to your heirs) if a careless accident while working on a device costs your life! For this reason, you *must* learn and consistently observe a few sound rules of technical safety before attempting to work on any appliance.

SAFETY RULES

Safety is really nothing more than a matter of common sense care. The rules of safety have been designed to prevent accidental injury.

General practices in professional machine shops and those followed by electrical workers are good ones for the home do-it-yourselfer. The safety tips listed in this chapter should be reviewed frequently and observed at all times while working on all appliances.

There are two basic reasons for observing the rules of safety. The first and most important is to protect your own safety. If you are careless and injure yourself, then you shouldn't be a do-it-yourselfer. Leave the job to someone else. You can replace the $50 you spend. You can even replace the entire appliance if need be. You *can't* replace an eye or a life.

Of secondary importance, but still important, is the safety of the appliance on which you are working. It doesn't do much good to save $30 or so in making the repair, only to waste $100 from damage you've done.

Quite often the two are interlinked. What protects you will often protect the appliance. This makes it doubly important to learn the rules of safety, and practice them until they become the natural way of working.

The following is a list of the most important factors for personal safety. Read them carefully. Better yet, read this section carefully and then post a copy of the list on page 34 in your work area and read that each time before you begin work.

1. Dress properly. Wear full-length pants covering your legs. No shorts, skirts, or clothes that have holes in them. A canvas lab apron can be used to cover your clothing if you wish to protect it, but it's best to work in old clothes that can be easily discarded if accidentally torn or stained. The complete covering from pants or a lab apron or overalls is to protect you from burns or shock while soldering or using electric equipment.

It is safer to wear rubber-soled shoes or to stand on a rubber pad when working with electricity. *Never* go barefoot, stand on concrete, or on the ground outside while working with anything electrical. Also, never work with electricity in an area where water has spilled on the floor or while standing on damp earth.

Protect your eyes. It's a good idea to wear shatter-proof glasses or safety goggles when soldering or working with motors that spin at high speed. Specks of grit or metal or hot solder flicked into an eye can permanently blind a worker.

2. Remove all jewelry. Metal jewelry is an excellent conductor of electricity. Keeping it on while working with electricity can lead to a harmful or fatal shock. Also, jewelry and loose belts or a tie can quickly become caught in moving machinery or gears when you are involved in repair work. This can lead to the loss of limbs, or loss of life if the equipment is big and powerful enough.

3. Follow the one-hand rule. Just as current flows from one side of a wire to the other, it can flow through your body. If both your hands come into contact with the circuit, the current flows through the body — across the chest and heart regions — which is very dangerous. Always keep one hand in your pocket or behind your back when measuring current or voltage. Some workers will use a metallic band around the wrist of the working hand, with a wire constantly attached from that band to electrical ground. Then, if a finger, knuckle, or elbow accidentally touches a hot line, the current runs through the shortest path to ground — through the hand or arm, to the wristband and its connecting wire to ground.

4. No food or drink. Do not work while consuming food or liquids. Spilled liquid becomes a short circuit path that can damage equipment, and spilled liquid or food can become a corrosive agent that can damage the appliance sometime in the future. Having food or drink nearby while you are working is also a distraction. It's more difficult to concentrate on probing that wire if you have a soft drink at hand. If you're hungry or thirsty, it's time to take a break. Go in the house, and have your snack while enjoying some television (or while reading the applicable sections in this book).

Never, *never* consume alcoholic beverages while working around electricity or power tools! Alcohol is a depressant. Your reactions and coordination are both lessened. Even a single beer can reduce your abilities enough to create a danger, and often you won't realize it until it's too late.

5. Keep the work area clean. The floors and desk or table top in your work area must be kept clean and free of any litter that might cause someone to slip, trip, or stumble. Some litter, if metallic, could even cause shock or a fire-generating spark if it hits "live" electrical circuits. A buildup of dust and dirt can make the job more difficult. It can also get into the appliance and create a number of future problems.

6. Have the proper work attitude. Don't clown around or engage in horseplay while using tools or working with power-driven devices. Many painful injuries are caused by the careless and thoughtless antics of a would-be comedian. Do not talk to or distract anyone when they are working with electrical devices or with potentially dangerous tools. And don't let anyone do it to you. If your long lost brother walks in while you're pulling out a heating element, stop what you're doing and give him your full attention. Or ignore him and give the appliance your full attention. You can't safely have it both ways.

7. Use the correct tool. Each tool has a particular function. Don't attempt to hammer with the handle of a screwdriver or use the point of a knife to turn a screw. Slips in attempting to use the wrong tool for a job are a major cause of cuts and bruises in workshops. And do not use *any* of your tools if they are not in proper working condition. Additionally, it is important to use the right *size* tool for the job at hand. It might be tempting to remove a screw with the closest screwdriver, regardless of blade size. Don't.

8. Cut away from the body. When using any type of cutting tool, make sure to slice or cut *away* from your body. If you slip, about the worst that will happen is that you'll damage the thing being cut. But at least you won't find yourself sitting there at the work bench trying to figure out how to get a knife out of your leg.

9. Take care of injuries immediately! Take competent care of any injury that does occur at once. Even the slightest cut or burn can develop serious complications if not properly treated in time.

No workshop, and certainly no home, should be without a complete first aid kit. Everyone in your home should also know how to use the kit. A first aid kit is all but useless if no one knows how to use it. (Worse yet are those homes that have a first aid kit and no one even knows where it is!)

Every area in the country has classes available in first aid, up to and including CPR courses. These classes are either free or very inexpensive.

An accidental electric shock that stops your heart need not be fatal — *if* someone in your home knows how to handle the emergency.

10. Unplug! Before you begin working on an appliance, be sure that the plug has been removed from the outlet. It's not good enough just to shut off the switch. There will be times when you'll have to have power applied to carry out a test. That's fine. In this case, your conscious and deliberate action will be to plug in the appliance. After the test is done, unplugging it again should be such an automatic reaction that you hardly realize that you've done it.

Many a finger has been lost by repair persons who took apart a disconnected fan to effect a repair, fixed the fan motor, plugged it in to test it, turned it off, then forgot to unplug the appliance before putting it back together. Accidentally hitting the power switch, the blades can swiftly become a human meat slicer!

Unplugging also applies to power tools. Never leave a soldering iron or a piece of electrical equipment under repair while it is still turned on or plugged in to a wall outlet. Don't leave the work area for any reason until you have double checked to make sure that all heat sources have been disconnected and that all potentially dangerous items have been put away in their proper place.

Young children are adept at finding an opportunity to play with dangerous power tools or around electrical appliances that may contain short-circuits or mechanical malfunctions. If you can't fix it right away, put it away!

11. When in doubt — DON'T. The line voltage that powers the appliance is dangerous. Even more dangerous is overconfidence. If a situation comes up and you don't really know how to handle it, stop! It's time to think things over. And if you can't figure it out, call in a professional.

Safety Rules

1. Dress properly	*6. Keep the work area clean*
2. Remove jewelry	*7. Use the correct tool*
3. Use the one-hand rule	*8. Cut away from the body*
4. No food or drink	*9. Take care of injuries*
5. Have the proper work attitude	*10. Unplug!*
	11. When in doubt — DON'T

DANGER SPOTS

Most of the dangerous spots in an appliance are obvious. By now you realize that anywhere that ac line voltage is present is a spot to be avoided. Avoiding anything hot is another obvious danger. Then there are the various mechanical dangers, such as sharp metal edges, screws, and fasteners, and various parts that may suddenly snap or shatter. However, not all danger spots are quite as obvious.

If one of the ac lines has been damaged in some way, it could be touching the metal chassis. This can also happen if you accidentally (or purposely!) cause a short between the incoming line voltage and the chassis. In essence, this could make the entire appliance a dangerous ac source. You can't see if voltage is present. The simple solution is to unplug the appliance before you work on it.

Heating elements are used in many appliances. Even those that don't actually have a heating element can have parts that get hot during operation. Many people make the same mistake with heat as they do with electricity. They expect to be able to see it. Although it is sometimes visible, such as when a heating coil glows, it isn't always. If you can safely do so, and without actually touching anything, move your hand over the top of any suspected hot components. Heat rises. If heat is present you should be able to feel it.

There is another danger that many people never think about, until they've experienced that danger first hand. This danger involves capacitors, or "cans" (so-called because that's just what they look like). Capacitors are electronic devices which can be used in several ways. One use is to separate or "filter" different types of current—ac and dc. A capacitor blocks dc, while allowing ac to pass easily. If a particular appliance is prone to RF (radio frequency) interference, capacitors can be used to pull that RF signal away from the appliance and to a ground. These capacitors are usually very small and present no danger.

The capacitor can also serve as a sort of temporary battery, storing a charge until it is needed. If the appliance has some sort of electronic circuit board with transistors or IC chips, then there will be a power supply for the device, which converts wall outlet ac to the dc needed to operate these solid-state devices. Capacitors are a major component of power supplies and are used to smooth the voltage fluctuation. The partially converted incoming voltage flows into the capacitor where it is stored and then released in a steadier flow. This works fine as long as the power is flowing; and if everything is working properly, the capacitor will drain itself of charge through a resistor after the power has been shut off.

Another use of a capacitor also takes advantage of the capacitor's

FIG. 2-1 A large capacitor can hold a dangerous charge.

ability to hold a charge until it is needed. Electric motors and compressors draw large amounts of current when they first start up. This extra drain presents two problems. First, the fuse or circuit breaker could blow. Second, since a relatively insufficient amount of current is flowing at startup, the motor or compressor can't do its job as well and will wear itself out more quickly. A startup capacitor solves these problems by providing a boost.

However, any capacitor can hold a charge for a long time, long after power has been turned off or after a power cord has been disconnected from the wall. Make it a point to short out large value capacitors with an insulated screwdriver (after the power cord is disconnected), before attempting to work on the appliance. To short out the capacitor, place the metal shaft of the screwdriver against one terminal of the capacitor and simultaneously touch the point of the screwdriver to ground, and then to repeat the process at the other terminal. Any charge left will short circuit harmlessly to ground, instead of later jumping through your hand to your chest.

There is another danger well worth mentioning here—a fire hazard in *every* home which few are aware of, and which few take preventive maintenance steps to avoid. Household dust, especially when it becomes moist, can be a very effective conductor of electricity. When dry it can be explosively flammable.

All wall outlets, light switches, and overhead and wall light fixtures are mounted in small metal or plastic boxes called electrical cases. These boxes are screwed or nailed to the studs or framework of the house, and then the outlet, switch, or light fixture is mounted over the case. The exposed (stripped of insulation) wire is wrapped around screw taps or solder connections, providing an electrical path to the outlet. Over periods of months and years, the outlet boxes become filled with dust. Some dust particles are *flammable*, as well as conductive. When an appliance is plugged in or out or a switch turned on or off, there can be a slight electrical spark when current first reaches the contact. This is caused when the contact is slightly corroded or otherwise not as solid as it should be. The spark that results is the *number one cause* of household fires in the United States. The most frequent occurrence is when an iron or a space heater, drawing heavy current, is pulled from a wall outlet or plugged into an outlet while the appliance is *on*. A spark naturally occurs when electricity jumps to the circuit being completed or opened by insertion or withdrawal of the plug.

Never plug an appliance in or remove a wall plug unless the power switch for that appliance has been turned to the off position. And it's a good habit once a year to remove all plastic wall plates covering outlets, switches and fixtures. Carefully use a plastic (insulated) extension on a vacuum cleaner to dust out the outlet boxes. If the vacuum attachment won't reach into the small areas, blow hard to get loose dust out, or use a small air compressor to blow out the dirt and grime.

APPLIANCE SAFETY

After you've made sure that *you* are safe, you can consider the safety of the appliance. All that is really needed is common sense.

If the appliance you are lifting is heavy, protect yourself and the appliance by getting help. Even then, be sure that you and your partner are using the proper lifting technique. Lift with the legs and not with the back. If you're bent over, you're not doing it right. The correct motion is an almost straight up-down motion, with the force being supplied by your legs.

If you feel the appliance slipping from your grip, don't be too proud to admit to it (immediately!) and set the appliance down again for a better grip, or a rest, or to simply abandon the attempt until even more help can be secured.

FIG. 2-2 Look for hidden screws and fasteners.

One of the greatest difficulties that both the novice and the old pro run into is that of removing or replacing the fasteners. Screw heads strip, making it difficult or impossible to move them. Worse yet, a screw or bolt can break.

In replacing those fasteners, many people have difficulty in getting the screw into the hole. So it goes in at a strange angle and is forced to cut new threads. This is *not* a valid function for a screw. Take your time, use the right tool for the job, and do it right. The day could come when you'll want to remove that screw again. (You also don't want to inadvertently have the screw cut into a wire.)

Some appliances have hidden fasteners. Others use plastic clips to hold things in place. With both, it's easy to damage the appliance — sometimes beyond repair — if you're not careful.

If the part you're trying to take off doesn't seem to want to move, don't force it. Look carefully for something else that's holding it. (See Chapter 4 for more information on disassembly and assembly.)

Chapter 3
Fundamentals of Electricity

To better protect yourself and the appliance, and to make the job easier, you should have at least a basic understanding of what electricity is, and how it does what it does. This is not a difficult or complicated subject. Most of it is merely a matter of understanding the terminology and learning the definitions of the words used. (To help with this, a glossary is provided at the end of this book.)

It's important to know the terminology. Without it you might be able to fix the appliance well enough, but you won't be able to communicate very well should you need parts or the advice of a professional. ("I need one of those round things that screws into that square-kinda thing over in the lower back corner by that long skinny thing.")

Knowing a little about how electricity makes things work can also help you in other ways. First and most important, when you understand how electricity does the job, you are less likely to be injured by it. Second, your understanding can also make it much easier to diagnose and handle malfunctions.

The main thing to keep in mind is that electricity always requires a complete path to work. (It's surprising how often even seasoned professionals forget this simple rule.) This is true for both ac and dc.

Think of a battery, such as the one you use in a flashlight. It has two sides. One is flat (the negative side), and the other has a small bump sticking up from it (positive). If you connect just one side of the battery to a device, nothing is going to happen.

The electricity often passes through a number of wires and components. If a break happens anywhere in that path, things can come to a grinding halt. At the very least, the way things operate in the appliance is going to change.

If everything stops working, nearly every time you'll find the problem in just one of two areas. Either the power isn't getting to the device in the first place, or there is a break in the pathway. This break could be something simple like a broken wire or blown fuse. It could also be from a faulty component that won't allow current to pass.

FROM THE SERVICE ENTRANCE IN

Power is carried along the heavy wires of the power company, through the large transformer that changes the primary voltage into the voltage needed in the home, and then through the meter and into the service entrance of your home. This is the main breaker or fuse box. There may also be one or more secondary boxes.

Everything up to the main breaker box is in the realm of the power company. Under no circumstances touch anything there. If there is a problem on that side of your service entrance, call the power company immediately and let them handle the rest.

From the service entrance and into the house is your responsibility. Local ordinances might make it illegal for you to work on the wiring, but

FIG. 3-1 The service entrance. Note the circuit breakers.

that just means that you must pay the cost of having a licensed electrician come out.

Total service amperage for most home wiring circuits is 100 to 200 amps, depending upon local housing codes. Most of the fuses or circuit breakers are 15 to 20 amp, and they control those circuits involving other household wiring such as to lights and wall outlets. Heavier 20- to 40-amp fuses or breakers are used for the circuits feeding 240-volt devices such as stoves, air conditioners, clothes dryers, etc.

The service entrance is always supplied with a main switch or main breaker, which can disconnect electrical service to the entire house. It can also be used to cut off the power supply from the utility company to the breaker box so that circuits can be rewired or new circuits added.

BRANCH CIRCUITS

From the main box, the incoming power is divided (distributed) across a number of individual circuits. As mentioned above, these circuits are controlled, and protected, by fuses or circuit breakers. Usually wires connect together a number of outlets or lights. You might have both lights and outlets on a single circuit controlled by the same breaker.

It's important that you label these branch circuits at the service entrance. If they're not already labeled, it's going to take some time and effort on your part to figure out what goes where.

Do this by a process of elimination. One way is to flip off a breaker and find out what outlets or fixtures are no longer powered. The other way is to shut all the breakers off and then flip them back on again one at a time to find out what *is* powered. This will also involve the use of a VOM to test wall outlets.

Expect to spend the better part of a day at this, and lots of running in and out. A piece of masking tape stuck on the cover plate of each outlet after it shows power both eliminates that outlet from future testing and gives you a place to write a number or other code. As the job of tracking everything down continues, it then becomes easier and easier since there are fewer places to test.

If there are 30 outlets in your home, the first test through might eliminate 5 of them. Mark those and the next run through the house with VOM in hand will give you only 25 to test. And so on until all have been marked.

Leave the lights switched on throughout the testings until they have been eliminated. This way you'll know which breakers operate which lights and which switches.

You'll probably find that there is a logical order to the way things are wired in your home. An outlet in the kitchen is unlikely to be powered by the same breaker as one in the garage, for example. A room with multiple outlets and lights (which will be almost all rooms in your home) will probably not run all from the same breaker.

FUSES AND CIRCUIT BREAKERS

Despite all precautions, it's possible for a short circuit or other malfunction to occur in an appliance. If this causes a heavy flow of current, the wires in the wall can heat up and eventually melt. A fire could start. After this you won't just lose the appliance — you could lose your home, or worse.

Fuses and circuit breakers are safety devices to prevent overloads and to keep short circuits from causing wires to overheat to the point where they become fire hazards. Fuses are rated according to the amount

FIG. 3-2 A fuse and a circuit breaker.

of current (amps) they can carry. When a circuit attempts to draw more current than the rated fuse value, it will burn out.

A circuit breaker is much like an automatic switch. It is a modern replacement for old style fuses. Like a fuse, it will automatically turn off whenever a circuit overload or short circuit tries to draw more current than an electrical device is supposed to use. Unlike a fuse, the circuit breaker needs only to be reset to return electricity to the circuit.

CHECKING FOR OVERLOAD

When too many appliances or lights are placed on a single electrical circuit the result is an overload. For example, a 15-amp circuit may be used to operate wall outlets in the living room and a hallway of a home. If lights in use are drawing a total of 7 amps, a TV set is on drawing one-half an amp, and someone then plugs in an electric iron that draws 10 amps while heating — that's a total of 17½ amps on a 15-amp circuit. The circuit will be overloaded. Some of the lights will have to be turned off in order to use the iron there, or the iron will have to be moved to another room or plugged into an outlet from another less-used circuit in order to prevent the overload.

A blown or burned out fuse indicates that you have placed an overload on that circuit, or that some part of the wiring within the house or within an appliance plugged into one of the wall outlets has developed a short circuit. Visual evidence is readily available for a circuit where the safety device has tripped. If a plug-type fuse has blown, and you can see the open fuse strip through the small window, it is generally an indication of a circuit overload. If the fuse burned out due to a great deal of heat, blackening the window, it usually means a short circuit has occurred.

Cartridge-type fuses must be checked for continuity on the resistance (ohms) scale of your VOM. Remove the fuse from its holder and probe it with your meter. If the meter shows no resistance (continuity) from one end to the other, it is still good. If it shows infinite resistance (an open circuit), the cartridge fuse is blown and must be replaced.

With blown screw-in plug-type fuses, a 25-watt light bulb may be used to test a circuit. When the fuse blows, turn off the main or master power switch in the service box. Unscrew the blown fuse and replace it with the 25-watt bulb. Put the main switch back into the on position. If the bulb burns quite dimly, there is an overload and some appliance or light must be disconnected before the circuit can again work properly. If the bulb burns brightly, there is a short circuit. Go back into the house and disconnect lights and appliances one at a time, going back to the fuse box each time to check the condition of the bulb. When unplugging an appliance or lamp causes the bulb to go out, you have found the unit containing the short circuit. Leave that appliance disconnected, turn off the main power, unscrew the light bulb and replace it with a new fuse. Turn the main switch back on, and take the faulty appliance to your workbench for a checkup.

With circuit breakers, there is either a "red flag" (a red piece of metal that shows through a window when a breaker is tripped) or the breaker switch will be moved over towards the "off" position. (With some breakers this movement is small.) To reset, just move the toggle switch all the way to the off position, then back to the on position.

No matter what type of protection is used in the wiring circuits of your home, try to determine what caused the fuse or breaker to go before replacing or resetting. Disconnect some of the lights or appliances on that circuit, in case it was an overload. If the circuit then works fine when reset or with a replacement fuse, you might — one at a time — turn on each of the lights or appliances that were active when the first overload occurred. This will help you determine if some of the appliances might have to be moved to another home circuit.

If the new fuse continues to blow or the breaker keeps tripping even after all appliances, lamps, and ceiling lights have been turned off, the problem is probably a short circuit in the wiring itself. You will either have to get professional help from an electrician, or use your VOM to try tracing down that short circuit so you can fix it yourself.

If the breaker (or fuse) holds with all lights off and all appliances disconnected, then — one at a time — turn on the light fixtures. If the safety device blows when you turn on one of the lights, the short is within that lamp or its on/off circuit or its power cord or plug — or within the bulb itself. To effect a repair, see the Lamp servicing tips in Section Two of this book.

If the breaker holds for all the lights, then carefully plug in and then turn on each of the appliances on the circuit. This way, you can discover which appliance was at fault and can begin repair work on that particular appliance.

In some instances, everything might work fine after the breaker has been reset or the fuse replaced. The short or overload which caused the trip could have been a temporary malfunction or a power surge from the utility company. This is particularly possible in periods of electrical storms, or when excessive electrical use by all consumers in a service area is causing lights to dim or flicker.

A note of extreme caution. *Never* replace a fuse with a new one of a higher amperage rating, or by using a penny or piece of metal as a temporary replacement. This is a highly dangerous fire hazard. Always use replacements of the exact same rating. Do not attempt to change fuses or work on a breaker box in the dark, or when standing on wet ground. Keep a flashlight handy on top of or beside the service entrance box, and stand on a dry, rubber mat if it has been raining or the ground is damp. It is a wise precaution to keep one hand in your pocket when working with fuses or breaker boxes. Throw the main power switch into the off position before working with individual circuits, and have someone else around to assist with first aid or in turning off the power in case of an accident.

HOME WIRING

In this section on electrical wiring, you will find a basic review of wiring circuits, a description of many of the terms used, working hints when handling electrical circuits and wiring, and a review of some specific problems and their cure. Working with electrical appliances is a lot easier, and a lot cleaner, than most mechanical work. But it is more dangerous if you become careless or rushed. In a few communities in this country, only professional electricians are authorized to do *any* repair or installation of home wiring circuits. In most areas, though, it's permissible for the home handyman to do the electrical work and then have it inspected by local authorities.

If any rewiring is to be done, whether the local code calls for an inspection or not, the work must conform to standard wiring codes and practices. Some of these rules are listed in this chapter. You should ask your local building authority (inspector) for a copy of the local code before undertaking any chores involving home wiring. Another good source of information is the National Electrical Code. It costs $15.00, and a copy can be obtained by writing to the National Fire Protection Association, Inc., Batterymarch Park, Quincy, MA 02269.

Even if local regulations make it illegal for you to tackle the wiring in your home, the home handyperson can, of course, work on any home appliance without calling in inspectors or checking with outside authorities. Often, however, the addition of a new appliance to a home necessitates a change in the wiring circuits from the fuse or breaker box or the installation of more wall outlets.

Occasionally, it is necessary to provide a 240-volt outlet (instead of the normal 120-volt outlet) for a window air conditioner or some other high-wattage (high power consumption) appliance. In these instances, you can either hire a journeyman electrician or electrical contractor or you can do it yourself following the guidelines of this book. If you do it yourself, get an inspection by the city building authorities after you have completed the work.

Again, check the local codes. In most instances you must obtain a building or an electrical work permit before starting the job.

TYPES OF WIRE

The type of wire used will depend upon your local electrical code and on where and how the wire will be used. For home wiring, the most commonly used is Type T. This is a single solid wire coated with plastic insulation. If the wire is to be in a wet area, then Type TM is used; if heat is present, Type TMH is used.

An easy way to run multiple wires is to use a cable assembly. Inside the cable are several wires, each with its own insulation. The designation on these cables is usually given as a number, such as 12-3 (3 wires of 12 gauge) or 14-2 (2 wires of 14 gauge).

Some electricians use metal-armored cable, commonly called BX cable but more correctly called Type AC. BX cable cannot be used except in areas that are always dry. The Romex insulation is either rubber or a heavy plastic coating. For use outside or in areas likely to become wet, building codes require that Romex be installed through a sealed conduit

Common Wire Types

TYPE	USE/COMMENTS
AC	Commonly called BX; metal-armored cable
HPD	Asbestos covered wire; heat resistant
NM	Multi-wire cable
NMC	Same as NM, but moisture- and corrosion-resistant
S	Flexible cord, with stranded wire
SO	Same as S, but oil-resistant
SP	Flexible cord, rubber-coated
SPT	Flexible cord, such as used for small appliances
T	Solid wire with plastic insulation
TW	Same as T but moisture-resistant
THW	Same as T but moisture- and heat-resistant

or pipe (either galvanized metal or PVC plastic piping). In any household work, it is best to use the same type of cable already in use for similar circuits in your home. Special grounding circuits are necessary when connecting Romex to BX, or BX to Romex circuits.

The size (gauge) of wire to be used depends upon the current it will be required to carry. The *smaller* the gauge number, the *larger* the wire. Eight-gauge wire is more than ⅛ of an inch in diameter, while 14-gauge wire is less than $\frac{1}{32}$ of an inch thick. The electrical cord supplying voltage to most appliances is either 12, 14 or 16 gauge. Irons, microwave ovens, toasters, broilers and units using more power (those containing heating elements) draw more current and need a larger (lower gauge number) power cord than electric razors, lamps, or radios that draw relatively small amounts of current.

Most home circuits work off a 15- or 20-amp fuse or circuit breaker. The wire for such a circuit must be of at least 12 gauge. If you are installing a new wall outlet to be used as a power source for a microwave oven or for a small room heater or air conditioner, it is best to use 10- or 12-gauge Romex or larger, since the appliance may use almost all of the 15 amps from the house fuse or circuit breaker, especially when first turned on. It is best to isolate these circuits to only one or two outlets, since an overload will result if you try to operate too many other appliances on the same circuit that feeds a high-wattage device.

For installation of wiring for a heavy-duty air conditioner or heater, consult local codes and experts. It may be necessary to install a circuit of a higher amperage for correct operation—for instance, a circuit breaker or fused circuit of 20 or 30 amps instead of the normal standard 15 amps. For this, wiring codes definitely specify 10-gauge or larger Romex feeding from the circuit box to the wall outlet. Use of too small a wire will lead to overheating of the Romex, and a potential fire hazard. This is what led the National Fire Underwriters Board to develop electrical codes. At times the specifics required of an installation might seem petty, but don't ignore them.

Electrical codes specify not just the size of the wire but also the color of the insulation. This is to help protect the people working with those wires. (Imagine opening a box and seeing a collection of colored wires, with no idea of what any of them are carrying.)

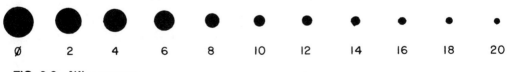

FIG. 3-3 Wire gauges.

As has been mentioned, a complete path is needed for electricity to work. This means that one wire brings the current to the outlet, and another takes it back. The one bringing in the current is the "hot" wire; the one taking it back is called the return, the ground, or the neutral.

The grounded wire is always white (assuming that the installer followed code). This wire can never be switched, fused, run through a circuit breaker, or interrupted in any way. The only time you won't find a grounded wire is when the outlet is supplying current for something operating only at 240 volts.

The "hot" wire for ac voltage is usually black. By code it can be anything other than white or green in color (and certainly can never be a bare wire).

Most modern wiring circuits use different metallic colorings. The "hot" side is generally considered to be the copper or brass-colored contact. The white-colored contacts, which are sometimes coated with nickel or tin to become silvery, are used for the white wires, or the ground side of the circuit. (For a socket, the white wire goes to the screw shell,

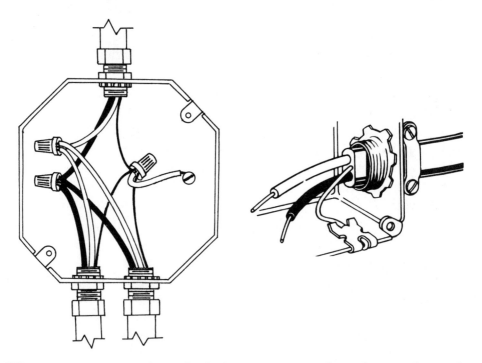

FIG. 3-4 How a junction box is hooked up. One wire is "hot"; the second is used as a neutral or return and is connected to the ground of the power company. The third wire is connected directly to ground to carry away voltage. Note that the "hot" wire is black. Reprinted by permission from *Chilton's Home Wiring and Lighting Guide* by L. Donald Meyers (Radnor, Pa.: Chilton, 1980).

with the "hot" connected to the center contact of that socket.) A screw contact that is green is the grounding wire, sometimes called the safety ground. The wire to it will be green, green with yellow stripes, or can be a bare wire.

It might sound confusing, but there is a distinct difference between the grounded wire and the grounding wire. The grounded wire does carry current, but it carries it back to the service entrance and into the power company's system. The grounding wire—to the green contact, and what is often a bare wire—doesn't carry any current at all during normal operation. Instead it serves as a safety and pulls the current away from the outlet or appliance and literally into the ground. Besides being used to pull the dangerous current away from an outlet, it is also directly connected to the metal chassis of an appliance. Then if something goes wrong, the metal body of the chassis or its parts (such as a motor) cannot hide a deadly charge because it has been discharged by the grounding wire.

All this assumes. of course, that the wiring has been done correctly. It should also give you a good idea as to why color coding, and paying attention to the codes in general, is so important. Open an electrical box for ac voltage, see a black wire, and you *know* that it's deadly unless you stop the power outside the house.

SWITCHES

Switches are lever, push-button, or revolving rheostat-type devices to make or break an electrical line; that is, they open or close a wiring circuit in order to turn an appliance on or off. With the circuit *closed*, the switch is on and the circuit is energized. An *open* circuit results when the switch is in the off position, and no current flows to the appliance.

Most wall switches in a home are of the single-pole, single-throw type (SPST). This means the switch is in use across only one line and either on or off. It is connected in series with the hot wire of a circuit (the black wire according to U.S. electrical wiring codes, although any color other than white or green may be used).

A double-pole, single-throw (DPST) switch has four contacts. It splits both the hot and the ground wires of a circuit. One side of the switch is in series with the white, or ground wire; the other side of the switch opens or closes the hot or black wire. The DPST opens or closes both wires when it is operated.

In cases where two switches are used to control a single circuit (an inside and an outside switch for a porch light or a garage door-opener), a

FIG. 3-5 A typical wall switch.

three-way switch is necessary. This type of dual control requires a three-wire cable between the two switches and the device to be operated.

In addition to wire and switches, there are outlet sockets, outlet boxes, switch or junction boxes, and connectors to secure the Romex or BX cable to the boxes.

Switch boxes, rectangular in shape, are made of galvanized metal or plastic and are used as the housing for switches or wall outlet plugs.

Junction boxes are octagonal in shape and are used for joining (splicing) wires from different circuits, or for mounting fixtures such as a ceiling light.

The outlets are either one-, two-, or four-plug sockets, designed to provide electricity when an appliance is plugged in.

For new installations, you will also need solderless connectors, which are twist insulated caps that "screw" over the ends of two stripped wires that have been twisted together (spliced), electrical tape, and either straps or insulated staples to hold cable to wall and studs as it is being strung from breaker box to final location.

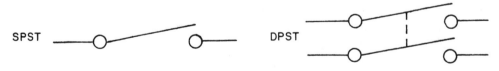

FIG. 3-6 Schematics of SPST and DPST switches.

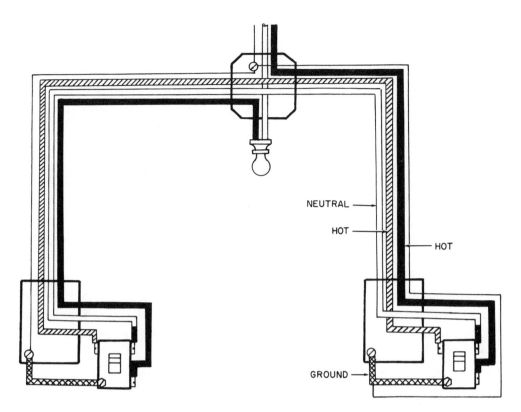

FIG. 3-7 Wiring diagram for a 3-way switch.

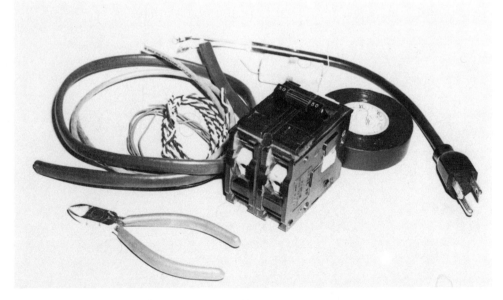

FIG. 3-8 Electrical supplies. From left, wire cutters, wire and cords, fuses and breakers, and electrical tape.

TESTING AND REPLACING ELECTRICAL OUTLETS

It's likely that sooner or later you will need to replace an outlet. Over the years, inserting and removing a plug from the outlet will wear it out. Beyond that, and despite everyone's efforts at quality (hah!), things do go bad. If no power is coming to the outlet, yet the fuse or breaker and the wires are good, then it's time to replace the outlet.

Testing an outlet is easy. You need nothing more than your VOM, set to read in the 120 vac range. Since you're working with ac, it doesn't matter which probe you use for which hole. Just make sure that you hold the probes *only* by the insulated handles.

First, put a probe into each of the two flat slots. The meter should show you a reading very near 117 volts. Next, put one probe into the larger of the two flat slots, and the other probe into the round hole beneath. There should be no reading at all. If you *do* get a reading, the outlet or wiring has either been hooked up wrong in the first place, or has shorted. Finally, touch one probe to that same round hole and move the other probe to the second flat slot. You should once again get a reading of 117 volts ac.

This test shows that the narrower slot is bringing the needed ac to the outlet, that the ground*ed* wire is carrying it back and completing the circuit, and that the ground*ing* (the rounded hole) is functioning as a safety.

It is recommended that you carry out the test several times, especially if you're not 100% comfortable in using the meter.

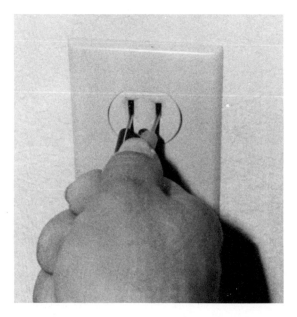

FIG. 3-9 To test a wall outlet, start by probing the two flat slots. The reading should be 117 volts.

FIG. 3-10 Next put one probe into the round hole and the other into each of the two flat slots. The reading between the round hole and small slot should be 117; between the round hole and the large slot, the reading should be zero.

The 117-volt reading is an average. It will fluctuate. A reading of anything from 104 to 127 is considered normal. A difference of a few volts won't matter. However, you may get consistently high or low readings every time you test. This condition might require correction. Contact the power company.

Some appliances will not easily tolerate changes in voltage. For example, a microwave oven will not cook properly if the supply voltage is too high. If the wall outlet used to feed your microwave consistently shows an output voltage of 125 to 127 volts, you might try operating the oven with a short (six foot or less) 10- or 12-gauge extension cord. This is a violation of a standard warning not to operate heavy-duty appliances with extension cords, and it is a poor solution. (You should call the power company.) But when the voltage supply is too high, the cord can provide the necessary voltage drop to bring the supply to the appliance down to the 117 or 120 volts needed to kick the microwave klystron into operation.

Motors and heaters will work inefficiently if the supply voltage is too low, and a low voltage will shorten the working life of these appliances. If the supply from the outlet to be used is consistently below 110 volts, then select another circuit for that appliance. It may be necessary to operate that particular appliance from a single (isolated) circuit, that is a dedicated circuit with no other electrical device plugged into it.

Replacing the outlet is easy. All you need is a screwdriver, although a knife or wire stripper might be needed if the end of the wire has been sparking and is charred.

First the most important step. SHUT OFF THE POWER AT THE FUSE OR BREAKER! If you're not sure which one it is (all fuses and breakers should be labeled at the service entrance), shut down all power coming into the house.

Remove the two screws that hold the cover plate and then lift that plate out of the way. Inside you'll see the outlet, the wires going to it, and two more screws that hold the outlet to the box. Remove these two screws.

You can now pull the outlet from the box. There is bound to be some stiffness to the wires, but don't yank on the outlet. A firm pull will get it out far enough.

Check once more that all power has been shut down. Although using a meter is fine for this, don't trust it. Go to the service entrance and double check that you have the power cut.

The wires should be color coded, as specified above. If you're uncertain, simply get out some masking tape and label the wires. If necessary,

FIG. 3-11 Changing an outlet.

clip off the ends (as little as possible) and strip back insulation to bare clean, shiny wire.

All that's left now is to reverse the procedure. Put the wires in place on the new outlet, tighten the terminal screws completely so that no bare wire shows (which will mean bending a loop in the direction of the turn of the screw if you've had to clip the wire), carefully push the outlet into the box, tighten those two screws, put the cover plate on, and tighten those two screws.

Before you consider the job done, get out the meter and test the new outlet completely, both for power coming in and for the proper safety grounding.

SOLDERING

To do soldering correctly takes practice. And the place for that practice is *not* inside the appliance. If you've never worked with a soldering iron or gun before, practice first on some scrap wire. Learn how to do it properly before you attempt it where it will count.

Before you begin the job, be sure that the wires to be soldered are perfectly clean. If they're not, sooner or later you're going to have trouble. Clip off enough of the end of the wire so that the wire is clean and shiny.

When making a splice or tap, be sure that the joint is physically strong. Solder is meant to create a joint that will conduct easily and that will resist corrosion. It is *not* meant to weld or glue the wires together.

Hold the soldering tool beneath the joint. Once the joint is properly heated, it will melt the solder. Resist the temptation to melt the solder with the soldering tool. If you do this you'll end up with what is called a cold joint. This means that the solder has not properly adhered to the wire, and you'll eventually have trouble with that joint.

Soldering Tips

1. *All wires must be perfectly clean.*

2. *Make the joint physically secure before you solder.*

3. *Apply heat to the joint, not to the solder. Let the joint melt the solder.*

4. *Use the solder sparingly.*

5. *Test the joint for continuity.*

If you look at the soldering work of a pro, you'll see that very little solder is used. There will be just enough there to "tin" the wires. You should never see blobs or drips of solder. If you've followed rule #3, the wires will melt the solder and it will flow evenly over the wires, turning them a shiny silver color. As soon as this happens, the job is done. Don't keep adding solder.

To make the job easier and more professional, take a few extra minutes and pre-tin the wires to be joined. This gives the wires a light coating of solder and will help to create a more secure joint.

REPAIRING WIRING AND CONNECTIONS

Not long ago we moved into a new home. It was then we discovered that our kitten had at one time or another decided that the lamp cord hidden beneath the couch was a wonderful chew-toy. Somehow the kitten survived and became a cat. But it meant that the lamp had to be rewired.

A number of things can happen that necessitate replacing a wire, cable, or power cord. Tugging at a sharp bend in the cord may have caused it to partially or completely break at the point of the bend. Rubbing action could have worn away or frayed the insulation, leaving wire ex-

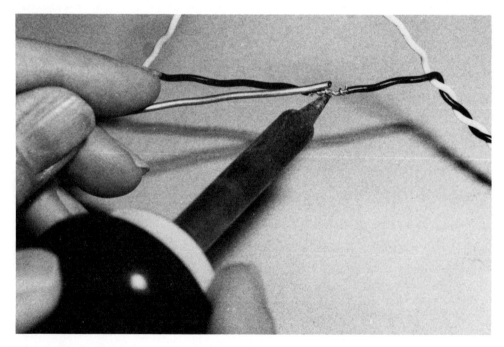

FIG. 3-12 Learn how to solder correctly.

posed. In some instances, the exposed wiring could have led to a short circuit, causing the electricity to travel instantly from source back to ground. The jumping over of electricity from one point in a wire to another leads to an instantaneous demand for high current from the power source (a wall outlet), and can easily generate enough heat to burn a wire in half.

Another potential problem is that someone may have stepped on the plug and damaged it, or that one lead of the two-wire cord may have been pulled loose either at the plug or as the cord enters the appliance.

You, too, might have a kitten chew through a wire, forcing you to replace it (and possibly the cat).

When you need to replace a faulty power cord for an appliance, here are the recommended sizes:

1. For clothes dryers, stoves, and large air conditioners, use 3-wire (grounded) 6- or 8-gauge cable or power cords. Use two separate fuses or breakers (or a double breaker) for a 240-volt appliance, protecting each of the two black wires (the two hot sides of the line). The white or green wire is the neutral line, or the center tap of the 240-volt source, and it should go to the ground terminal of the plug and the ground terminal of the appliance.

2. For small kitchen appliances (toasters, broilers, mixers, blenders, etc.) use 10- or 12-gauge wire or cable. It is best to operate frequently used kitchen appliances off of a 20-amp circuit in order to prevent overloads.

3. For lamps, radios, TVs, and movable appliances such as the vacuum cleaner, replacement cords should be 2-wire 12- or 14-gauge. Electric water heaters need to be serviced by a 2-wire 12-gauge cable, on a 20-ampere circuit.

The size of wire used for replacement cords is not all that important unless that appliance is drawing a lot of current. The easiest way is simply to take the old cord (or a piece of it) along and match it for size with the new wire or cable that you buy.

Many appliances and power tools come from the manufacturer with extremely short cords. The idea is that the owner is less likely to slice through the cord, or to knock over the appliance, if the cord is short. However, in some cases the cord is so short that you have to use an extension cord just to be able to use the appliance or tool at all. Or you have to replace that supplied cord with a longer one. (It is *never* wise to attempt to operate irons, space heaters, stoves, microwave ovens, automatic irons (mangles), or air conditioners with an extension cord. The cord may overheat — a fire hazard — and a resultant low supply voltage may shorten the working life of the appliance.)

When replacing a cord with a longer one, or when using or repairing an extension cord, you must be careful not to go too long a distance. Voltage is lost if a wire is of too small a gauge over too long a run. If the wire fails to deliver the needed voltage to an appliance or power tool, not only will the device fail to work at top performance, but there is a danger that any electric motor and certain other components will burn out quickly.

There are two "most important" considerations involved in the repair and installation of wiring. First, the ends of the wires involved have to be bright and clean before they are connected. Copper wire should be shiny and bright. If it is dull or blackish in appearance, either thoroughly clean the wire until is is bright and shiny, or trim back farther until you get shiny wire. If need be, replace the whole wire. Settling for less than a perfectly clean wire is dangerous.

Second, the final connection must be physically solid and insulated. This can be accomplished by using screw clamps, soldered connections or solderless caps, and by covering the connection with electrical tape. For splices and taps, learn the proper methods for securing the wires together. Do this before you complete the job by soldering the joint.

Never work on an electrical appliance while it is plugged into a power outlet. Disconnect the plug, and place the defective unit on a table or workbench. If the cord has been broken completely in two, the proper splicing procedure is simple (see below).

SPLICES AND TAPS

Joining the ends of two separate wires is called a splice. When a wire is joined at right angles to another continuous wire, it is called a tap.

The first step is to remove insulation from the wire about 1 to 1½ inches from the end. When using a knife for this removal, cut a slant as if you were sharpening a pencil, but be careful not to cut or even nick the wire under the insulation. With wire strippers, insert wire into the hole or opening of the correct gauge, close the strippers and pull the insulation off.

After peeling back the insulation, use your soldering iron or torch and some resin core solder to tin the exposed surface of the wire. This is not actually necessary with solid wire, but it is important with wire that is stranded. The tinning process does two things. First, it holds the strands together. Second, it provides a secure and permanently clean end. (Copper will eventually corrode. Solder won't.) The end result makes for a better electrical contact in the finished joint.

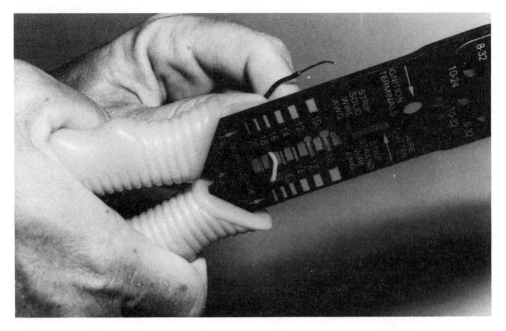

FIG. 3-13 Using a wire stripper.

Heat the wire with the iron, and apply solder. (Do *not* heat the solder to let it drip onto the wire. The wire, not the soldering tool, should melt the solder.) Shake off any excess, leaving the exposed end shiny and bright. If you've done the job correctly, the tinning will look like nothing more than a change of color. You shouldn't be able to see any blobs or other obvious signs of the soldering.

To make a tap joint along a continuous wire, peel away about 2 inches of insulation from the wire to be tapped, and about 1½ inches from the end of the connecting wire. Securely wrap the end of the tap wire around the continuous wire and carefully solder the connection. Then wrap the entire joint with electrical tape to reinsulate the connection, leaving three wires coming from the one joint.

To splice the ends of two separate wires, strip the insulation, pre-tin, and then twist the ends to be connected tightly. A solderless cap may be screwed on, or the twisted wires may be soldered and then covered with electrical tape.

However you finally join the wires (solder, tape, etc.), the splice or tap itself must be physically strong. A gentle tug before completing the job will tell you. Don't assume that the solder is strong enough to hold the wires together. The only time you don't need to worry about this is when you are using a solderless screw cap, since this automatically twists the wires together for a secure splice.

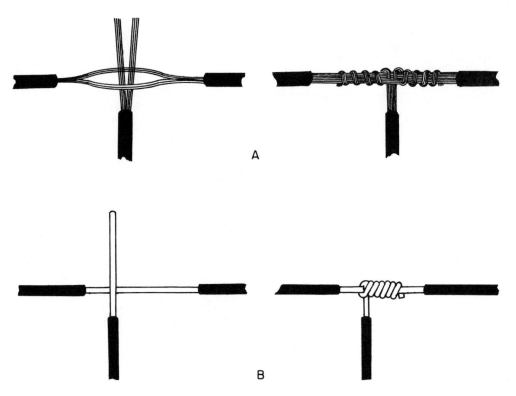

FIG. 3-14 A. Wire tap method for standard wire: *left*, intertwine wires; *right*, twist tightly to make a physically secure tap. B. Wire tap method for solid wire; *left*, strip the end of a wire, and the spot to be tapped; *right*, twist the tapping wire slightly to secure.

To fasten the wire to a screw terminal, bend the end of the wire into a hook or loop. Attach the hook *in the direction the screw will turn when tightened*. This will hold the wire tightly under the head of the screw. If the loop is inserted the other way, tightening the screw may cause the wire to slip out from under the head.

In instances where the cord was broken close to the appliance end instead of the plug end, use the screwdriver, nut driver, pliers, or Phillips head driver to remove any screws holding the case together. You must take off any plate or covering so that you can find where this other end of the cord is attached inside the appliance. Generally, one wire will connect to one side of the appliance on/off switch, and the other wire of the power cord will attach to another point (or terminal) inside the unit. Removing the short part of the damaged cord, stripping the wire ends, and reattaching the good longer part of the cord will give you practice in desoldering and resoldering electrical wiring.

To desolder the old connection, hold your soldering iron or gun in one hand and the desoldering tool in the other. Make sure the appliance

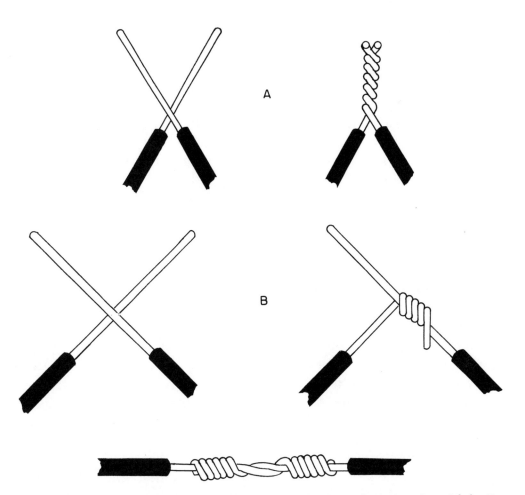

FIG. 3-15 Splicing methods. A. *Left*, remove insulation; *right*, twist wires tightly. B. *Left*, strip and cross wires; *right*, wrap downward on one side, then do the other side; *bottom*, completed joint is neat and strong.

is firmly clamped to the table or workbench, leaving both hands free to work on the unit without knocking it to the floor.

The soldering iron has to be held directly to the soldered contact point where the wire is attached long enough for the solder to completely melt. As the solder heats, it will become shiny and fluid. Place the sucking end of the desoldering tool directly over the terminal you are heating, with the iron still in place, and suck off the melted solder. You may have to repeat the process two or three times to get rid of all the solder and make the connection loose. Then use your needle-nose pliers to pull the wire loose from the terminal.

If the wire has been twisted tightly to attach it to the terminal before it was soldered, it may be necessary to melt the solder with the iron in one hand, and use the pliers in the other hand to wiggle and pull the wire loose from its connection. The appliance must be firmly clamped in place while the desoldering/disconnecting attempt is being made.

When the break or insulation damage is somewhere near the middle of the cord, the wire ends should be stripped and tinned (lightly coated with solder). Apply the soldering iron directly to the exposed end of the wire, and then apply a length of resin-core solder to the side of the wire away from the iron and hold in place until the solder begins to flux (melt). When the wire begins to take on the shiny, silvery appearance of a coated piece of metal, take away the iron and shake off any excess solder. This pre-tinning process will make it easier to solder the wire end to any other connecting point, and will serve as a binding agent to keep the individual strands of twisted wire from fraying and causing an accidental short.

After pre-tinning all four wire tips — two on each cord end (six tips if the cord is for a grounded, three-prong electrical connection) — take one of the wires from the longest end and twist it together with an end from the shorter piece of cord. In case the wires of the cord are covered with insulation of different colors, match the colors you are twisting together — black to black, white to white, green to green, red to red, etc. If the wires are of the same color, then use the VOM to check for continuity between one end of the wire still attached to the appliance and to the chassis or case of the appliance. This wire should go to the wire on the plug end of the cord that attaches to the widest prong. In two-wire electrical outlets, the wide prong, feeding to the wider left-hand opening of the wall outlet, is considered to be the ground side of the power cord.

Take your soldering iron or gun and apply heat and resin-core solder to the twisted wire connection. After the soldered connection has cooled, use wire cutters or diagonals to trim off any excess lengths of wire extending beyond the soldered joint. Then take your roll of electrical tape and wind a length around the spliced connection, making sure not to leave any wire exposed. The tape is to replace the insulation peeled off with a knife before the splice was made. It is not necessary to wind a lot of tape around the soldered connection, since the insulating quality of electrical tape is very good. Two or three layers of tape should be enough.

Repeat the process with the second wire from both the long and short end of the cord, again matching colors of the wire or wire insulation. The trick is to be sure that no point of the soldered connection on one wire makes physical contact with an exposed wire from other parts of the splice. If you have a three-wire connection, splice the third ground after you have used tape to reinsulate your second soldered joint.

Use the VOM to check the quality of your splice after completion.

You should show continuity from one prong to one wire at the other end of the cord, and no continuity or an open circuit from that wire end to any other prong.

PLUGS

When the break in the cord has occurred very close to the plug, or at the other end of the cord near the place where it enters the actual appliance, you can discard the shorter part of the cord and reattach the longer portion.

On most modern equipment, the plug is solidly attached to the end of the wire and cannot be removed and resoldered. In this case, purchase a standard replacement plug from any hardware store or supermarket. Dress the end of wire from the longer cord at the place where it was broken; that is, make the exposed ending clean and neat. Use a knife or wire cutters to clip off the damaged end neatly. Make the cut at a slight angle or on the diagonal (about 30 to 45 degrees).

Most general replacement plugs are the slip-on and clamp, no-soldering type. There is a clip on the side of some plugs that has to be opened. Push the cut end of the cord firmly into the new plug, and close the clamp firmly. Inside the clamping part of the new plug are two needle-like contacts that will puncture the insulation on the ends of the cut wire and establish a connection between each wire and one of the prongs of the plug.

Heavier duty appliances (which need higher values of current to operate) will use cords with thicker wire that can carry this current without overheating. Many of these cords cannot use the slip-on type of replacement plug, but will require a connector that has screws and a mounting plate for the attachment of the cord. In this case, you will have to strip the ends of the wire before attaching the plug. That means you'll have to remove about 1 inch of the insulation from the end of each wire in the cord.

If you have a solid rubber or plastic-coated cord that seems to be only one wire, use a sharp knife to carefully peel away the outer layer of insulation. This will expose the two (and often three) small wires inside the cord, each individually wrapped again with insulation. The three-wire cords are for appliances that must be grounded for safe operation. This means that the case or chassis of the appliance must be electrically connected with earth ground in order to prevent accidental shocks to the operator. Such appliances should only be used when plugged into a three-hole, electrically grounded outlet. The replacement plug should

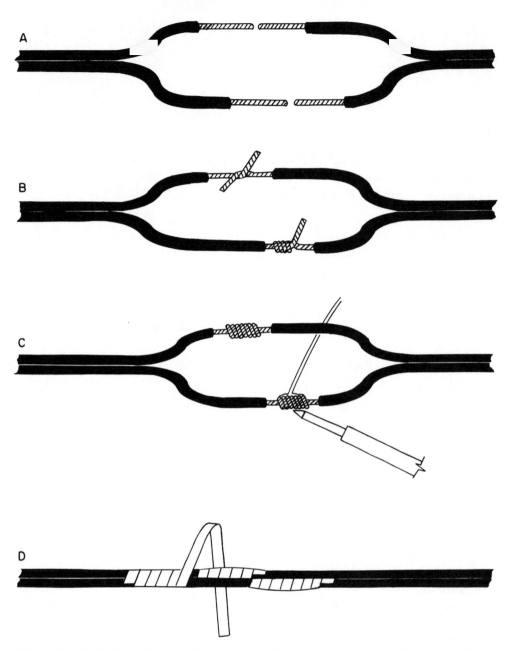

FIG. 3-16 A double splice. A. Remove insulation and stagger the wire ends. B. Wrap wires tightly. C. Apply solder to both joins. D. Wrap the finished double join with insulating tape.

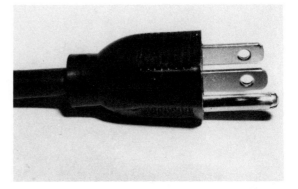

FIG. 3-17 This kind of plug is a permanent part of the cord. If the plug is bad, you'll have to cut it off and attach a new plug to the clipped cord.

be a three-prong plug. Figure 3-19 shows a three-wire cord, properly stripped for the attachment of a new plug.

Twist the exposed ends of each wire, making sure you keep the strands of one wire from touching or twisting with the strands from either of the other wires. (Tinning the ends will help.) Push the cord up through the hole at the end of the plug away from the prongs. Wrap each of the exposed wire ends around one of the three contact screws leading

FIG. 3-18 Electrical plugs.

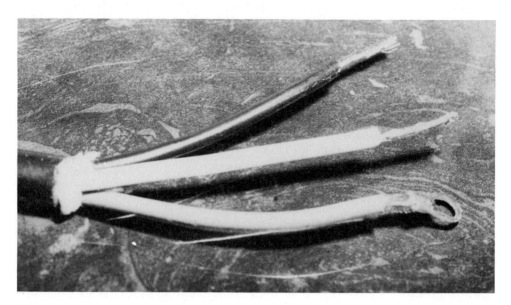

FIG. 3-19 A 3-wire cord, ready for its new plug.

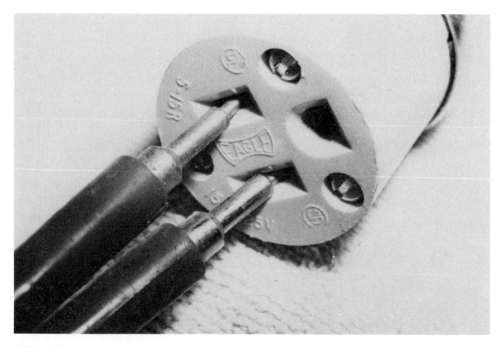

FIG. 3-20 Test the new plug and cord for continuity.

Chilton's Guide to Large Appliance Repair/Maintenance
FUNDAMENTALS OF ELECTRICITY

to the individual prongs. Again, make sure no frayed strands are left that may touch from one wire to another. Tighten the screws.

Replacement of this type of plug is best practiced on a throwaway extension cord or a piece of electrical cord that is open at both ends. This way, you can use the VOM to test how well you have done before trying to repair an actual appliance. If your work was sloppy or improper, the test meter will show you what has been done wrong. When you can attach replacement plugs that readily test correctly on the VOM, then you can expect to be able to properly repair ones broken on actual appliances.

Chapter 4
Large Appliance Repair Basics

Diagnostics and servicing, whether on a small appliance, large appliance, your car, home computer or transistor radio, is always about the same. It's a basic 1–2–3 approach.

1. Diagnose — Test the appliance to see what is wrong in general. Find the general symptoms (what works, what doesn't).

2. Pinpoint — Isolate the problem to the particular part causing the trouble.

3. Repair/Replace — After you've located the specific cause of the problem, repair or replace the bad part(s) and put the appliance back into service.

BASIC DIAGNOSIS

Diagnosis is nothing more than a process of elimination. There are only so many things that can go wrong. By eliminating those things that *are* functional, you'll find what *isn't*.

The diagnostic procedure begins with the simple and obvious, then goes to the more complex. Much of the time the problem will be simple and obvious, so why waste time and energy, and risk damaging the appliance by tearing it apart, when you don't have to?

Much of this process of elimination involves an "either-or" approach. For example, either power is getting to the appliance or it's not. In one quick and easy step you've identified either the appliance as the problem, or everything else coming to it.

At this point, if there is no power to the outlet, you'll know that taking the appliance apart will be a waste of time. The trouble probably isn't with the appliance. (If it is, the appliance has a major short circuit that is blowing the fuse or popping the breaker. And this in turn tells you something very important—you're probably looking for a bare or burned wire.)

If there is no power to the outlet, then either the problem is between the breaker and the outlet, or it's between the breaker and the power company's primary. This might sound oversimplified, but I knew a person who spent several hours ripping apart a nonfunctioning stove, and damaged his kitchen floor when sliding out the oven, only to find out that someone had crashed into a power pole and all the power in the neighborhood was out.

If you have power anywhere in the house but none at that outlet, then you already know that the problem is within a specific circuit. Replace the fuse in that circuit, or reset the breaker. Now try the outlet again. If the appliance goes out again, one more quick test will tell you whether the trouble is in the appliance (a short circuit) or in the wiring (also a short circuit). Reset the circuit again and try the same outlet with

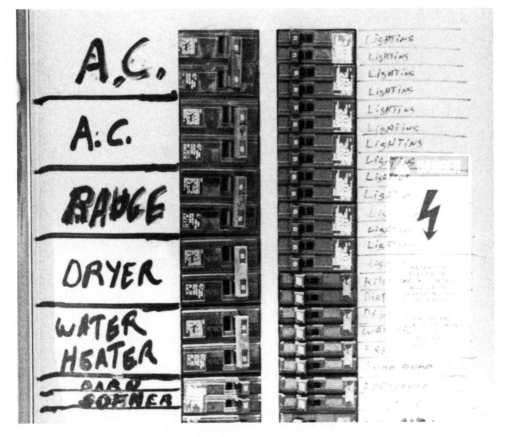

FIG. 4-1 Check the breaker (or fuse) and the outlet.

another appliance. (Also don't forget that the whole problem might be that you have too many things on that circuit. If the circuit is meant to supply a maximum of 15 amps, and you're trying to draw 20 amps from it, it will blow again and again until you remove the overload.)

Now if the circuit blows when you plug in your alternate appliance, you know that there is a short somewhere in the wiring between the breaker and the outlet. You can eliminate the breaker and the outlet itself easily. First visually inspect both (with the power off). You might see signs of a burned wire, which means that arcing has taken place. This is usually a sign of either a bad component or a loose wire. (Chapter 3 will give you more details on how to check/replace the outlet and fuse or breaker.)

If the circuit is not overloaded, and if only the suspected appliance is causing the breaker to let go, you should look for a short circuit in the appliance.

Short circuits, whether within the appliance or in the wiring of your home, must be located and repaired. This is both for proper diagnosis and repair of the appliance, and for safety. You will not be able to further test the device until you locate and cure that short, or until you can fix the device to the point where you can plug it in and turn it on without again blowing a fuse or tripping a breaker.

On the other hand, if there *is* power to the outlet, and if other appliances work just fine in that outlet, then you know that the appliance has indeed something wrong with it.

Within a couple of minutes you'll have eliminated at least half of the possibilities and can then concentrate on where the problem really is. If the problem is definitely in the household wiring, go back to Chapter 3. If it's in the appliance, continue reading.

The next step involves thinking and asking yourself a number of questions. Since you know that appliance, you probably know the answers to many of the critical questions that make diagnosis easier. Once again, begin with the simple and then go to the more complex.

Were you using the appliance when it went bad, or was the appliance working perfectly last time and failed to work the next time you went to use it? Has it ever worked? Has that particular function ever worked? Have you checked all the obvious things, such as the plug? (A large percentage of "malfunctions" are something obvious, or even silly, with nothing being wrong at all.) Does the person who first reported the malfunction really know how to operate the appliance?

A young couple bought themselves a brand new washer and dryer. The washer worked fine. The dryer wouldn't do anything. They called out the serviceman to find out what was going on, and nearly demanded

Initial Diagnostics Checklist

1. *Have I read the instructions?*
2. *Has the appliance ever worked? Has that function ever worked?*
3. *Is ANY part of the appliance working?*
4. *Is power getting to the outlet?*
5. *Is power getting to the appliance?*

that he bring along another dryer. That serviceman had been in the business for many years and knew that he could probably fix the dryer in a matter of minutes and that a replacement wouldn't be necessary. He guessed right. Within a couple of minutes he discovered that no one had read the instructions. They hadn't even read the labels on the buttons. Several buttons were labeled with various settings for the type of clothing being dried. One was labeled "On." The reason the dryer wouldn't work was because they expected it to work as soon as they pushed the button for the cloth setting.

If the appliance is new, you must answer "I don't know" to the question "Has it ever worked?" Even if it has been around for a while, it's possible that you're trying out a function that is new for you. Have you read the instructions? It could be that you're not doing the right things to get that function to operate. It's even possible that you're trying to get the appliance to do something it can't do.

If you've had the appliance for some time, and have used it successfully all that time, then it's less likely — but still possible — that you're doing something wrong. But you still can't assume that someone else (who says it's not working) knows how to operate it.

Especially when someone else told you that an appliance isn't working, you must confirm that the appliance *is* in fact nonfunctional. Then examine the unit while still trying to make it operate, so you can begin to diagnose the real problem. Now you begin to take note of the symptoms.

Did it overheat after long use and then malfunction? (After it's cool, if it seems to be working all right, it could indicate a lubrication problem, or electronic components that are wearing out.) Did a simple mechanical jar, made while moving the appliance from its normal working space to the shop, jiggle a loose part back into place?

If the appliance has a heating element that is failing to even warm

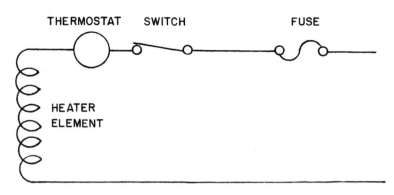

FIG. 4-2 Schematic of a heating element.

up, don't immediately jump to the conclusion that a thermostat or the heating element itself has gone bad. Begin the process of elimination. Think of the basic rule of electricity (it must have a complete path) and use your common sense. You'll know what parts to test.

With the heating element example, first there must be electricity available. At this point you should have already eliminated all things outside the appliance and up to the outlet. Now you can take one more step. Is power getting *to* the appliance? If that appliance has an "on" indicator and this fails to light, that is a sign that perhaps power is not getting to the appliance. The trouble could lie in the unit's power cord, power switch, or in any of the wiring or electrical controlling/switching devices that feed the heating element.

After you've determined that the fault is definitely with the appliance, it's time to find out just exactly what is wrong inside. As in the example above, the first part of diagnostics will help to eliminate many of the parts so you don't waste time on them.

PINPOINT AND ISOLATE THE FAULT

Step 2 in the diagnostics process is to pinpoint the problem. For this, you may have to take the appliance apart in order to test and visually inspect what is happening inside the casing. Most of the time, how to take the appliance apart is obvious. Some tips will be given in the next section of this chapter.

Whatever malfunction you are able to verify by testing the appliance for yourself, try to use common sense and your knowledge of the work normally accomplished by the device in order to think of all the wiring and parts that feed or that conduct electricity up to the part that is not doing its job. Use a notepad and jot down all the things you believe might be preventing the malfunctioning part from working.

If the motor doesn't turn, the actual problem might include power not getting to the outlet, a faulty power plug or cord, a faulty on/off switch or other function (such as speed) switch, an open wire inside the unit, motor binding due to lack of lubrication or a bad motor bearing, the windings of the motor itself might be shorted or open, or the blades or motor driveshaft could be physically bent and/or jammed.

Once again, go to the easy and obvious first. Use your senses. You might be able to see what is wrong. A wire might be burned, broken or loose. A belt may have fallen off. Look around carefully, both before you remove any panels or the casing, and afterwards. (Be sure that the power is off and the appliance unplugged before attempting to get inside!)

Continuing with the motor example, listen for a humming or vibration. If you hear it, you can probably eliminate (at least for now) those electrical things outside the motor. Some power is getting to the motor, or it wouldn't be making noise.

Shut off the power and see if you can find something mechanical at fault. Is the motor secure in its mount, or is it flopping loose? If the motor drives the appliance with belts, are the belts tight enough (but not too tight)? Are the belts even there? If the motor is gear driven, are the gears meshing properly? Are all the teeth of each gear intact?

Manually try to turn the shaft, if possible. Any binding should be fairly obvious. It should also be obvious if the binding involves the motor shaft, or whatever part of the appliance the motor is trying to spin.

If you keep the functions of the appliance and of the various parts in mind, you should be able to quickly and efficiently track down the source of the trouble.

The three overall functions of a household appliance are electrical, mechanical, and plumbing. A problem in the third category will generally be obvious, since it will usually leave a puddle of water.

Electrical problems are easy to track down by just keeping the "complete path" rule in mind. If electricity is getting to a certain component, but that component isn't working, it's highly likely that the component is at fault. If current isn't getting there in the first place, then the problem will be found ahead of that component.

To test a component for continuity, you'll usually have to remove the component. At very least you'll have to remove at least one of the wires (otherwise you may get a false reading and end up actually testing something elsewhere in the appliance).

A heating element, for example, should have a very low resistance reading if it is whole and sound. If it is starting to go, the resistance will be fairly high. If a wire within the element has burned through, the continuity testing will show an infinite resistance, indicating a broken path. If so, it is time to replace the element.

Mechanical problems can often be tested manually. A motor shaft can be turned. A button can be pushed. They can also be inspected visually for excessive wear, metal flakes, broken or loose parts, and an odd appearance in general.

Sometimes a component has more than one action. A button has a mechanical motion, and this movement may control something electrically. Pushing that button makes a physical, and consequently electrical, contact. In such cases, testing might include testing with a VOM and testing with your fingers.

DISASSEMBLY

In disassembly, your first concern will be to find out exactly how the appliance is put together physically. Even if you have an instruction or service manual for the device, most do not detail the correct way to get inside. Other books, and even the owner's manuals, might say a lot about what you can do once inside, but they may not tell you how to "break and enter" in the first place.

Keep in mind that any and all warranties might be voided with the first turn of the first screw.

Always switch off the power and unplug the unit. Even then, keep in mind that there might be charge-holding devices inside (such as capacitors) that can still be dangerous. Move carefully and keep your eyes open.

Now carefully inspect the case and the obvious holding screws. Many screws or bolts go inside the unit and hook up or fasten to something inside the case, too. Usually these are obvious, but not always. If you're in too much of a hurry when you take the appliance apart, a spring might pop loose and flip across the room. You could lose a part, and not be able to discover exactly where it was supposed to go anyway. Use care in removing any screw or clamp. As a general rule of thumb, screws toward the outer perimeter merely hold the case in place, while screws toward the center might be holding something else.

When in doubt, first slightly loosen the screws you *think* hold the case and nothing else. With the screws loose but not removed, you should be able to tell if something else is coming loose inside, or if just the case itself is being released.

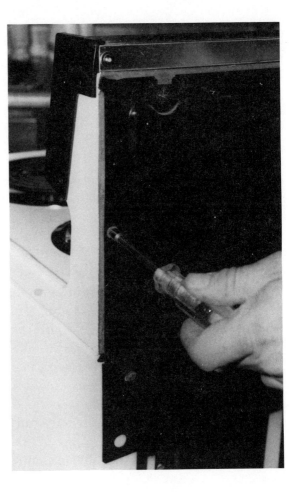

FIG. 4-3 Visually inspect the appliance cover(s) to find the screws, bolts, and other fasteners. Be careful not to remove those fasteners that hold internal parts.

Once all visible screw heads have been removed or loosened, carefully try to lift the covering plate or appliance case away from the unit. If it does not move, run your fingers around the edges to see where the cover or case is still attached. You may discover that you have to remove a function knob or a switch before the case can come off. *Never* force the case or cover off.

In these days of plastic mass manufacturing, many cases that appear to be sealed single pieces are actually snap-apart units. To find the "secret" release(s) is a matter of probing and prying.

If all else fails, and if the unit is a throwaway if you can't fix it yourself, then — *and only then* — apply force. Expect to physically break the case, and possibly destroy the appliance. At least you may find out how to open the next one without breaking it. There's a remote possibility that you will be able to fix the appliance, and then repair the broken case before reassembling.

In the disassembly process, use your notepad to draw diagrams of exactly what small parts go where, in what order they are mounted on a shaft, and what color of wire goes to what plug or connection.

Once the unit is apart, it is quite possible that you will find the malfunction has been caused by something easily repairable. A wire could have broken or could have pulled away from its terminal. If you are sure what terminal the loose wire is supposed to have connected to, then re-solder it in place.

Be very careful here after reassembly and plugging the appliance back into a wall outlet. If you soldered the loose wire to the *wrong* terminal, you may get a short circuit. Such a mistake could cause a major burnout and a major problem, when the original loose wire was a minor problem—if only you had hooked it back up in the right place. It can also be a potentially deadly mistake.

For this reason, if you find a wire disconnected inside and are not absolutely sure (either from diagrams in the user's manual or from the broken remains of the connection on the terminal), then consult a service person at the shop where you bought the appliance to be sure "point X" is precisely where the loose wire is supposed to be soldered.

(There are some very rare instances when a loose wire inside is supposed to be loose. The manufacturer needed a two-wire cable to hook up part A to part B, but was able to mass-purchase a bunch of three-wire cables at a warehouse fire sale somewhere. The maker would use two of the wires in the three-wire cable to manufacture the appliance, clipping

FIG. 4-4 Once inside, carefully inspect for any signs of obvious damage.

or taping the spare wire somewhere out of place. Taking the unit out of its case may have loosened that unused third wire — so don't just guess that it is supposed to be resoldered to some point. Check with a technician to be sure.)

The VOM can be a major help in your pinpoint work. You can use the resistance scale to test continuity or to see if a connection where the solder looks old and cracked is still making a good electrical contact. Sometimes using a VOM to test what appears to be a bad component will reveal that it does *not* need to be replaced.

At this point, if you have still not found the broken part or wire in the appliance, place it carefully on the workbench so that no exposed wires are touching each other or making contact with outside metallic surfaces. Plug the unit into a wall outlet, turn the power switch on, and use the voltage range (ac voltage) to be sure the electrical pressure is getting to the switch or to a malfunctioning motor or heating element.

If the appliance uses a dc motor instead of an ac motor, it will have a power supply circuit board for internally converting the wall outlet ac to dc. The trouble could be in this power supply. (See the next section in this chapter.)

Remember to set the VOM to its highest voltage range before applying power, and then reduce the range switch if necessary to read the voltage accurately. For most home service work, however, measuring exact voltage values will not be necessary. If voltage pressure of any value shows on the meter when the probes are placed across two points that should indicate a voltage, then the circuit is probably working all right. For this

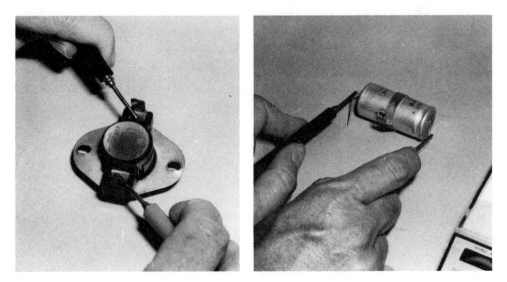

FIG. 4-5 Using a VOM to test the continuity of a component.

reason, some shop people use an inexpensive ac/dc testing light bulb instead of a VOM for voltage checks.

The diagnostic and pinpointing steps involve a lot of physical inspection and some solid mental work. "What *could* be wrong to cause the unit to do this?" Why waste your time testing the motor when the symptoms show a problem with a heating element only? (Remember the old adage "If it's not broken, don't fix it.")

With over half of your malfunctioning small appliances, the pinpoint steps will lead you definitely to the broken or bad part. You will be able to *see* what has gone wrong, and will be able to either repair, resolder, or replace the defective component. Once you have done this, reassembled the appliance, and then tested it to see that it is again working properly, you have achieved your first money-saving, quality appliance repair job!

If you've reached the point where you *think* you know which part is bad (because everything else seems to be okay), then it's best to read further into the chapters on individual parts testing before spending bucks for a new part (that may actually not be required). If the part checks good, you'll have to repeat these preliminary diagnose/pinpoint steps, and will probably discover that you missed something. You assumed a part or wire connection was working, and it really wasn't.

POWER SUPPLIES

Some appliances use microchips and other electronic circuitry that requires voltage other than what comes from the wall outlet. If your appliance is "computerized," this is definitely the case.

To operate, those circuits generally need dc voltage of between 5 and 24 volts. Since this doesn't come from the wall outlet, something must be built into the appliance to convert that 120 vac into the needed value of dc. This something is a power supply.

There are three basic parts to a power supply: the transformer, the rectifier, and the filter(s).

The transformer takes the incoming 120 vac and drops it to near the final value(s) needed. This happens because of a difference in the number of windings between the primary and the secondary sides of the transformer. If that ratio is, for example, 10:1, then the 120 vac will be dropped to 12 vac. (This is an oversimplification, but gives you the general idea.)

The size of the transformer depends on how much current it has to handle. If the components needing dc voltage (or even a different value

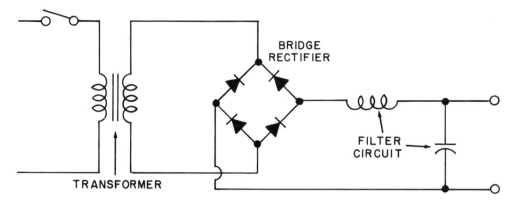

FIG. 4-6 Schematic of a simple power supply.

of ac voltage) draw a large current, then the transformer will be large. If very little current is required, the transformer can be small.

After the voltage has been changed to the proper value, a rectifier is used to convert the ac into dc. A rectifier allows current to flow in one direction only, while blocking the flow in the other. This is done with a component called a diode.

The usual rectifier contains multiple diodes. If just one diode is used, half of the incoming current passes through; the other half is "thrown away." Current flows only half of the time. By using two diodes, when one diode is blocking half of the flow, the other diode takes over and passes it along, still in the same, single direction. Many power

FIG. 4-7 A power supply transformer.

FIG. 4-8 A bridge rectifier.

supplies use what is called a bridge rectifier, which is four diodes connected in a square pattern. Most often this bridge rectifier is a single block.

Once the current comes out of the rectifier, it is direct current only in the respect that it flows in only one direction. It still varies in value from zero to (for example) 12.

A filter capacitor takes care of the rest. As the current flows into the capacitor, it fills up and stores the energy. It then releases it in a steady flow. The end result is virtually pure dc of a steady value that can be used by the other components.

Testing a power supply is easy. First find where the ac comes in. Set your VOM to read ac in the proper range (usually the 120 vac line voltage) and see if ac is getting to the power supply. If it is, find where voltage leaves the power supply. Most of the time, the power supply output will be dc at between 3 and 24 volts. Set your meter to read at least this much and test the output(s).

If power is getting *to* the supply but is not coming out, the power supply is bad and has to be replaced. If it is coming out, or power isn't getting to the power supply in the first place, then the problem is elsewhere.

BUYING PARTS

In many instances, you will discover that the problem with a nonfunctioning appliance can be cured by cleaning, lubricating, or by physically bending or twisting a misshapen part or bracket back into its correct position. To effect repairs in other cases, you will find it necessary to purchase a replacement part. Sometimes parts availability is a major problem, but luckily this is seldom a troublesome factor in appliance work — unless the make or model of appliance you are working on has been discontinued by the manufacturer quite a few years before.

Once you have determined what replacement part is required, make a note of the brand name and model number of the appliance. This is generally listed on a plate somewhere on the outside of the unit.

Either go to the store where the device was initially purchased, or consult the Yellow Pages of the telephone book in order to obtain the name of the main dealer in your area for that brand of appliances. The service shop of that store will have the part on hand or should be able to order it for you.

Also listed in the Yellow Pages are likely to be stores that carry appliance parts. (After all, the professional service people have to go somewhere to get their parts.) Chances are very good that you'll find more than one in any medium-size city.

Some appliance components are available at general hardware, electrical, or electronic supply outlets. In most cases, if a part is generally available, it is less expensive to pick it up from the counter of one of these stores than to order the replacement from the original manufacturer.

Occasionally you will run across an electric motor, a heating element, a thermostatic control device, or some other commonly used component that is not stocked on the counter of a general supply store, but which may be ordered directly from the part manufacturer rather than from the maker of the appliance. The motor of a Brand XXX vacuum cleaner, for example, may have been supplied to the Brand XXX company by International Electric Company, with the IEC home address stamped right on the motor casing. Write for an IEC catalog, or call to see if there is a local dealer. Buying direct from IEC may again save you the normal mark-up charged by Brand XXX if you buy the replacement from that manufacturer.

Many mechanical parts that have broken or that have been damaged in other ways can be replaced by duplicates you fashion yourself in your home workshop. Or the part could be fashioned by a local machine/tool shop for you at a reasonable price if the component is a fairly simple one. This could prove necessary if the part is not available to buy and if the price of the custom-made part is less than the price of a new appliance.

It's an excellent idea to keep your old nonrepairable appliances as a stock of replacements for newer units when they, in turn, go bad. This is a good reason for sticking to the same brand and model of each particular appliance you use (once you find one that is completely satisfactory in its service).

For small appliances this is easy. If an old Mr. Coffee maker develops a bad switch and the unit is so old you'd really rather have a new one, don't throw away the bad coffee making machine. Some day your new Mr. Coffee unit may encounter a burnt-out heating element, and you'll have a quick repair part right out in the garage or workshop.

Larger appliances present more of a problem, simply because of their size. However, if that dryer is going to be trashed anyway, you can still scavenge it for parts and throw away only the bulky case and frame.

Even when switching model or brands in a new appliance purchase, some parts may still be common. You can disassemble the broken unit and store all the individual parts in a conveniently marked and labeled parts bin or storage cabinet. A switch, thermostat, bracket, or heating element may work as a future repair part not only for the new appliance you bought, but also in some other appliance that just happens to require the same type of component.

Chapter 5
Electric Motors

Almost every appliance in and around the house has either a heating element or a motor, or both. The common solution when a motor goes out is to replace it. However, there are some very good reasons for not using this as the *only* solution.

First, you may be able to restore the motor to operating condition — often quickly and easily. What seems to be a major problem may be something simple; or perhaps the problem can be fixed by installing a new part (instead of a whole new motor).

Second, even if you do buy a new motor in order to get an appliance back into working condition, the old motor can serve some useful purposes. Unless the old motor has a trade-in value, take a few moments to break it down. You might find that the problem is something fixable, which will give you a spare motor for next time. If it can't be fixed, good parts in the old motor can be salvaged.

Perhaps most important is that the old motor, fixable or not, can serve as a valuable training aid. Understanding how a motor works and does its job in the appliance can help tremendously when it comes time to swap out a faulty motor.

MOTOR BASICS

A motor changes electrical energy into mechanical energy. To oversimplify: plug it in and it spins. The amount of force in that spinning is called torque. The power of a motor is generally given in horsepower. But whether rated or not, and no matter how the motor is rated, you should always try to get an exact match when replacing a motor. (Quite often, if the replacement motor isn't an exact match, it won't even fit.)

FIG. 5-1 A typical appliance motor.

In brief, motors work on the principle of magnetism: Like magnetic poles repel each other, while unlike poles attract. If you set up a number of magnets in a wheel with a ring of magnets around this, and if you could constantly shift the polarity from south to north to south, etc., you'd have a motor.

Fortunately, electromagnetic forces will do this for you. Electricity flowing in a wire generates a magnetic field, causing the wire to become a "magnet" temporarily. By using polarity switching (either mechanically or by using ac voltage, which constantly changes anyway), the motor spins merrily away to get the job done.

Motors have various types of voltage connections to their internal coil windings. *Series-wound* motors have field windings connected to the brushes of the armature so that current passes through the field windings before reaching the armature (mounted on the rotating shaft of the motor). These motors are used to move heavy loads that have to be started slowly. Series-wound dc motors must be operated always under load, to prevent speed from building up.

Shunt-wound motors have a parallel wiring setup to provide current to field windings independently from that supplied to the armature coils. Shunt-wound motors are used where constant speed is necessary under varying load conditions.

Compound-wound motors incorporate both series and shunt armature windings. These motors can handle heavy starting loads and can draw energy from the centrifugal force of a spinning flywheel during peak load conditions. The flywheel also absorbs extra energy from the motor after peak demands pass, helping to maintain speed of the motor and to increase the working life of the device.

The electric motor is designed to provide a turning force (torque) to accomplish some function designated. There are three main groupings of electric motors, each designed for a specific type of power source: direct current (dc), single-phase alternating current (ac), and polyphase alternating current (ac). The first and second are those commonly found in homes, while the third is more likely to be found in commercial applications.

AC MOTORS

There are a large number of ac motor types, each with special characteristics to meet the needs of various appliances. The most commonly used types of ac motors include: universal, shaded pole, capacitance start, split phase, PSC (permanent-split-capacitance), and synchronous. These motors all share the common trait of requiring ac to operate, with the exception that the universal motor can be designed to operate on direct current as well.

The universal motor is aptly named, since it can be used to handle

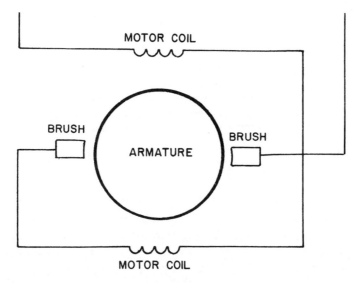

FIG. 5-2 Schematic of a universal motor.

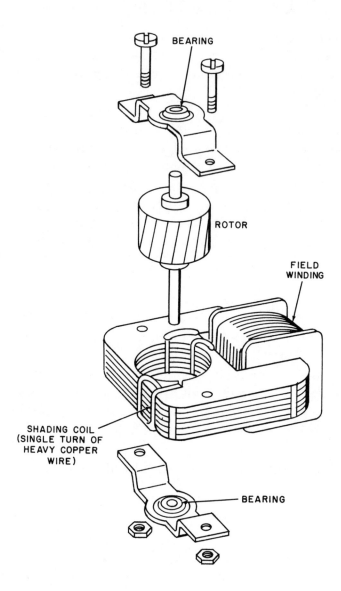

FIG. 5-3 Shaded pole motor. Reprinted by permission from *Reader's Digest Fix-It-Yourself Manual* (New York: Reader's Digest General Books, 1977).

so many different jobs and can operate on either ac or dc, depending on the way the appliance (and the motor) is set up.

Shaded-pole motors can be used when the amount of torque needed is relatively small. This motor has four stator poles, with a band of copper somewhere on the poles. The copper band "shades" the magnetism, thus causing the same pole to have more than one magnetic field, which

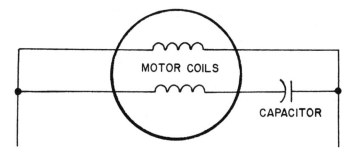

FIG. 5-4 Schematic of a capacitance start motor.

in turn causes that magnetic field to rotate, which in turn causes the motor to rotate.

Watch the lights in your house when a major appliance, such as a large air conditioner, starts up. They tend to dim for a moment. The reason is because it takes more current to get a motor started than it does to keep it going. This is hard on the electrical supply, and hard on the motor. Nicely enough, there is a solution. A capacitor stores energy. When the motor starts, the capacitor releases its energy, giving the motor the needed boost. Capacitance-start, split-phase, and PSC motors are often found in large appliances for this reason.

In split-phase motors, two windings are used instead of just one, with a centrifugal switch to shift from one to the other. The start-up winding brings the motor up to about 80% of operating speed. The centrifugal switch then cuts out the start-up winding and brings in the main winding. Then the remaining 20% of operating speed is gained and held.

PSC motors are generally found in larger appliances such as air conditioners and other devices requiring large amounts of torque. These motors are like a combination of capacitor start and split phase motors. In a PSC motor, the capacitor is permanently connected in series with the motor, which makes the start-up winding do double duty. At startup, this winding works in the same way as the startup winding of a split phase motor. As the motor reaches speed, the winding will work as the main winding.

Caution

In any motor that uses a capacitor, use caution in testing. The function of that capacitor is to hold a charge. This charge can be dangerous. Don't trust it, even if the capacitor seems to be very small.

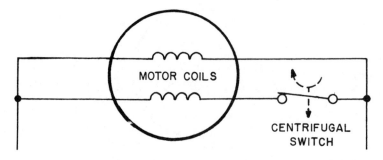

FIG. 5-5 Schematic of a split phase motor.

All ac motors are *synchronous*; that is, they are automatically synchronized to the incoming 60-cycle power. This causes them to run at a designed speed. (The speed of a dc motor is determined by the amount of current flowing, since dc has no frequency.) Appliances sold in the United States are designed to operate on the standard 60-cycle frequency supplied by power companies to the wall outlets of your home. The speed of a *hysterisis/synchronous* motor is dependent upon that 60-cycle frequency (and with the number of poles used in the manufacturing process).

For those interested, the basic formula is S = 120 x f/p, where S is the speed of the motor, f is the frequency of the line voltage (60 cps in the United States), and p is the number of poles in the motor.

The formula isn't exact. There are other forces applied against the motor, such as the voltage level and the load the motor is expected to handle. Most motors operate within about 5% of the speed resulting from the formula. And actual operating speed of most motors tends to vary across a fairly wide range.

There are times when this speed variation isn't sufficient for the job, and an exact motor speed is essential. It requires a specially built motor that is designed to operate at exact speed, and one that is relatively immune to variation.

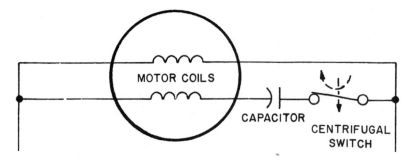

FIG. 5-6 Schematic of a permanent-split-capacitance motor.

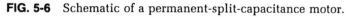

DC MOTORS

Audio and video cassette decks and other high-tech goodies, as well as battery-powered toys, often have dc motors. The most common form of dc motor uses an internal permanent magnet to provide the magnetic field in which the armature operates. Other forms of the dc motor use various methods of connecting the stator and armature to enable the motor to operate using a dc power source.

A convenient feature of dc motors is that the direction of rotation can be changed simply by changing the polarity of the current feeding it. This is how the motor direction is changed in order to reverse the movement of the device. The switching arrangement for the motor and connected device contains an internal polarity changing feed to the motor.

You will sometimes find compound-wound dc motors in fans and blowers and in heating and cooling household appliances. The speed may be controlled either by switches or by a rheostat that varies the revolutions per minute (rpm) over a wide range (by varying the current). The more common two- or three-speed manual switching arrangements incorporate a number of resistors that can be cut into or out of the fan/blower motor circuit at the discretion of the operator. When the switch is in a low-speed position, more resistance is added to the motor wiring circuit than when the intermediate speed is selected. Removing all resistance from the circuit by throwing the switch into another position puts the motor into high-speed operation.

RELATED PARTS

In air conditioning and other heavy-duty motor appliance operation, a magnetic clutch is often involved in order to adjust the driving power of the motor. You will also find these clutches on electric lawn mowers and in some vacuum cleaners. Magnetic clutches are generally in the form of an electromagnetic coil (solenoid) mounted between the drive shaft and the belt pulley.

Electric motors sometimes have accompanying starter voltage supplies, clutches, drive belts, gears, and operational solenoids—so many circuits and subcircuits that it becomes difficult for even professional service workers to keep abreast of new developments. Yet despite the involved circuitry and mechanical principles, only a few basic electrical/mechanical processes are involved in their operation. Once you have studied the basic operation of relays, solenoids, and electric motors, new adaptions are easy to handle when you encounter them.

In today's world of solid-state electronic improvements, many dc motors are directly mounted on a small circuit board that can contain a transistor or two, coils, capacitors, and even an IC (integrated circuit) chip. The electronic parts mounted on the board are part of the motor control circuit. They can help in determining the rotation speed of the shaft, they can be part of any designed overload protection devices, or they can supply the correct starting voltages when the unit is first turned on. Generally, the manufacturer will supply a replacement motor *only* as part and parcel of the entire mounting circuit board. In other words, the motor (a simple $10 dc unit) cannot be purchased unless the consumer buys a whole new motor board (for $25 to $100 or more).

In this case, you would be wise to look for the name of the motor manufacturer, as opposed to the appliance manufacturer. It would be less expensive to order the motor directly from the maker, without the mounting circuit board and components.

A difficulty arises due to modern "proprietary information laws." The motor could be a fairly common style made and marketed worldwide by ABC Motor Company. But when XYZ Appliance Company builds a trash compactor or an electric juicer using this motor, XYZ will contract ABC to supply 100,000 or so electric motors with a single variance from normal manufacturing specifications from all the other ABC motors. An internal capacitor or resistor may be added when ABC assembles the motors for XYZ. This means that if you buy a "stock" ABC motor as a replacement for an XYZ appliance, it won't work. And neither ABC nor XYZ will let you know what tiny modification has been made from one motor design to the other. In this case, you will have to order the replacement motor and the accompanying attached circuit board (which is still okay on your own appliance) directly from XYZ Appliance.

The fact that a motor in your appliance may be mounted on an independent motor circuit board means that other test steps should be taken before ordering a replacement motor or motor circuit board. Use the VOM to test continuity through the board. If you are familiar with basic transistor testing, remove leads and use your VOM (or a transistor checker) to see if an inexpensive transistor in the drive board is at fault, rather than the motor. Most important, use the voltage scale of the VOM to see that the dc value applied to the motor terminals is correct for proper operation (24 volts for a 24-vdc motor, etc.). If the voltage is abnormal, then the trouble is probably not with the motor. Instead, check individual components on the motor drive board, and test values from the circuit's power supply. (See the sections on testing and troubleshooting below.)

MOTOR MOUNTS AND BEARINGS

To successfully replace a motor, you have to know how to unmount the old motor and how to correctly remount the new one.

Some motors are designed to be held rigidly in place inside the appliance. Others are suspended on belts or rubber mounts, allowing them to vibrate rather freely. Be sure, when replacing a motor, that you allow it to be mounted the same way as the original. Don't attempt to clamp down a motor shaft that is supposed to be suspended loosely, and be sure to securely bolt or screw in place those motors intended to drive from rigid mounting brackets.

In efforts to keep manufacturing costs low and to maintain moderate prices for consumers, basic motor design for many low-priced appliances has changed a lot over the past decade. Many low- and moderate-speed dc motors are now constructed with plastic or silicone/plastic bearings.

A bearing is the small lubricated ball on which a motor shaft rotates. It eliminates friction as the motor spins to drive a pulley or belt or whatever. In old-fashioned kid's roller skates, it was easy to see the ring of metal ball bearings held in a row around a metal ring or collar. This collar was locked in place, with the wheel shaft protruding through it. The shaft would turn as the wheels rolled, rubbing against the freely rotating ball bearings.

At that time, most motor bearings were built in a similar manner. They would last for a long time, but required frequent lubrication.

FIG. 5-7 Testing the motor's bearings can usually be accomplished by turning and wiggling the shaft and feeling for excessive play or excessive resistance to movement.

Nowadays, in many sealed motor units, the bearings are plastic. They do not last as long. If the motor is a sealed, one-piece unit, when the bearings wear out the motor or the entire appliance is defective and ready for replacement. In such cases, you might as well do a little experimental test work. What you learn just might make such a motor salvageable, for use as a future replacement.

Whenever a motor using plastic bearings is *not* sealed, make it a point during routine maintenance checks to test the motor shaft's "freedom of movement." If the drive shaft wiggles too much when side-to-side pressure is applied, then the bearings are becoming worn. Exact replacements will be available, at low cost, directly from the manufacturer.

When an appliance is purchased, the retailer normally provides an operator's manual and a replacement parts list. Maintain a single file somewhere in the house for *all* your appliance manuals and repair information. If you order replacement bearings when you first notice a motor's rotation becoming "soft," then the new parts will probably arrive before the appliance suffers a complete breakdown.

Disassembly for replacement of bearings is fairly simple. Carefully study the way the unit was originally put together. Use the proper tool (screwdriver, allen wrench, nut-driver, etc.) to open the motor casing and remove the old bearings. Replace with the new part, being sure to align it exactly the same way as the original. Put the unit back together, simply reversing your disassembly steps.

Do not attempt to replace worn bearings with substitute "similar" bearings from your tool box. These fittings are critical for smooth, efficient, and trouble-free operation. Often the size variances are in thousandths, or even millionths, of an inch. What looks to be a bearing or bearing set of the same size could easily be wrong, and can damage the entire motor in mere minutes of operating time.

WHY DOES IT BREAK?

An electric motor usually becomes defective for just one of three reasons. Each *can* be repairable, but the time and expense involved generally means it's better to purchase a replacement. The three primary reasons a motor stops working are binding, an open circuit, or a short.

Binding is a mechanical problem. It is caused by any of several things. Lack of lubricant is a common problem, and it is not always easy to cure since many motors are sealed units. The bearings or bushings could be worn out, or they could be lacking lubricant, or both. It's also possible that physical damage has occurred. The shaft might be bent, or the pulley could be misaligned on the shaft.

Binding and other physical problems are detected by examination. There will probably be resistance when you try to turn the motor shaft, but it shouldn't be immovable, nor should it be uneven in its motion. (Also check to see if the shaft wiggles. This is a sign of worn bearings, even if the motor is not binding.)

Open circuits are most commonly brought about by a broken or burned wire or winding inside the motor. In some cases, this break might have occurred at or near the motor's terminal, making it easy to spot. When a wire is broken, current cannot flow and the motor stops. The solution for a break in the wire outside the terminal is to simply repair or replace the wires, or solder them back into place.

If the open (break) is in the windings, it is unlikely that you will be able to do much to correct the problem. However, if the motor is otherwise a delegate for the trash barrel, it might be worth the effort.

Gently remove the insulating covering of the motor windings to expose the coils of wire wrapped around a core. Unsolder the connection from the wire end on the outside of the winding and slowly begin unwinding it. Sometimes the break is on the first or second layer. Splice the connection in this manner:

1. With a knife or wire strippers, peel off the layer of lacquered insulation surrounding the ends of the broken winding.

2. Twist the two cleaned, pre-tinned ends together; resolder.

FIG. 5-8 Start your search with the easy and the obvious. Are the wires securely connected to the motor's terminals?

3. Cover the soldered splice carefully with insulating tape, to prevent the possible shorting out of some of the turns of the winding.

4. Rewind the wire, in as close to the original position as possible, and resolder the end to the connecting terminal to the voltage source.

Warning: This process could easily ruin the motor. Do *not* try it on a motor that has a trade-in value.

The third cause of motor malfunction is a short circuit. Somewhere inside the motor windings, the insulation has been scraped or burned off and a few or all of the winding turns have been shorted out. This means that the motor may try to turn, but it will either rotate slowly or will just sit in one position and hum or vibrate.

Most of the time, such a short circuit in the internal winding will lead to an arcing of the voltage from one exposed hot point to another, and the heat will burn the winding open. The cure, unfortunately, is usually to replace the motor.

As detailed above, a motor that has failed due to binding or other physical problems can be diagnosed by examination. The second two causes (opens and shorts) can be detected by some simple tests.

PROBLEM DIAGNOSIS

As a first step in diagnosing an appliance fault, you should understand exactly in what manner the appliance is failing. For instance, if the appliance has a plug, is it plugged in? This may sound terribly elementary, but it is an essential step in disciplined problem diagnosis. An example is the disposal found in the kitchen sink. Generally speaking, a disposal is often wired as an integral part of the home wiring system, but in some homes there is a little receptacle buried in the dark recesses under the sink, into which the disposal is connected. Putting items under the sink can some times crowd the space and knock the plug out of its ac receptacle.

If a motor has been under heavy load or running for a long time, it will sometimes just stop for no apparent reason. In a situation such as this, look for the thermal overload circuit breaker which, if present on the motor, is usually a little button on the end of the motor case, often colored red. This button should be pushed in, which resets the circuit breaker allowing the motor to run again.

These little heat protection circuit breakers can be found on a variety of motors, usually associated with heavier-duty appliances or elec-

FIG. 5-9 Some motors come with a built-in circuit breaker.

tric-powered tools, but they can be found on other appliances around your home.

The motor might directly drive the appliance. It might also drive it through the use of gears and/or belts. Visually inspect all of these parts for proper function.

Check all of the obvious things first. More than one person has replaced an expensive motor, only to find that the actual problem is elsewhere, and something very simple.

Pay attention to all of your senses. A burned winding will create either an open or a short, depending on where it is. Either way, the motor won't work. That burning will probably be visible if you can get inside the motor. Whether you can get inside or not, the burned insulation almost always gives off a sharp odor.

If the appliance is motor-powered, does the motor turn? If the motor doesn't turn, does it make a humming sound or other noise indicating that power is getting to the appliance? Is it getting too hot? (If the motor is getting overly warm, this is an indication of it driving a load that is too heavy, a lack of lubrication, or both.)

TESTING THE MOTOR

Once disconnected from a circuit, with any pulleys, drive belts, or drive shafts removed or unscrewed, the motor can be checked as an independent electrical device.

Check to see if the shaft is solidly in place. If it rotates freely but wiggles quite a bit from side to side, the bearings are worn out, and the only cure is to replace those bearings. Lubrication will not help. The only cure is to replace the bearings, or replace the entire motor.

Next, see if you can physically rotate the motor shaft with your fingers. If it is bound up, you may have frozen bearings, or overheating may have caused the metal to expand so much the shaft has bound somewhere inside the casing.

Especially with a permanently sealed motor, once it is frozen in this manner it is useless. If it has no trade-in value, you might as well take the chance of ruining the case and motor to see if you *can* get it operable again. The best way is to take a very small electric drill, smaller than ⅛ inch, and drill a hole through the casing by the bearings or other spots that might need lubrication. (Be VERY careful with this. Before proceeding, you should already have the motor apart so that you know exactly where to drill.)

Use a vise or clamp to hold the motor in place. Drill in an upward direction so any filings from the operation will fall outside the motor, instead of inside the casing and in the windings or bearings.

Once you have a hole in the casing, you can either use some type of high quality penetrating machine oil, or a silicon lubricant can be squeezed through the opening you've created. Allow the lubricant to soak for a few hours, then again try to turn the shaft by hand. Make sure you allow enough oil or silicon to settle in place so that the shaft spins freely and easily, and listen closely to see that there is no grinding or scraping noise from inside as the motor turns. If the motor is still noisy, try to lubricate it again. But keep in mind that over-lubrication can be almost as bad for a motor (and sometimes worse) than under-lubrication. Both can destroy the motor.

If the shaft stays in place and spins freely, the lubrication may have been all that was needed.

Check to see what type of voltage is required to operate the motor. An ac motor generally runs on 117-vac and can be tested by running test wires with alligator clip endings from the motor terminals to the voltage connectors on the appliance. When this has been done, carefully plug the appliance into a wall outlet and turn on the on/off switch. This way you can see if the motor is now operating correctly. If so, you can reassemble the appliance to see if your lubrication work was enough to get the unit working again.

If you have a dc motor, remember that without a load it can run for only a few seconds without going into high speed and self-destructing. *Never* apply ac to a dc motor. Always apply the correct value of voltage from a dc source, and for just enough time to see that you have freed the motor's binding enough for it to run. Then disconnect the motor, the wall plug of the appliance, and reassemble the unit.

TESTING WITH A VOM

A motor winding is nothing more than a coil of wire twisted around a core. The wires come out to a terminal strip or other connector. It's usually a simple matter to visually inspect those connectors to determine which contacts go to which of the internal windings. After you've determined which leads go where, set your VOM to read ohms in the x1 scale.

The two terminals of an individual winding should show resistive continuity. If the VOM indicates infinite resistance when placed across the terminals of a winding, there is an open somewhere in that winding. (If there is continuity between two different windings, there is a short between those windings.)

The reading for continuity should be very near zero. However, it should not *be* zero. The wire inside the motor should give a small reading. A reading of zero probably means that you are doing something wrong. It could also mean that there is a direct short between the two contacts.

Just as you should get continuity through a single winding (a read-

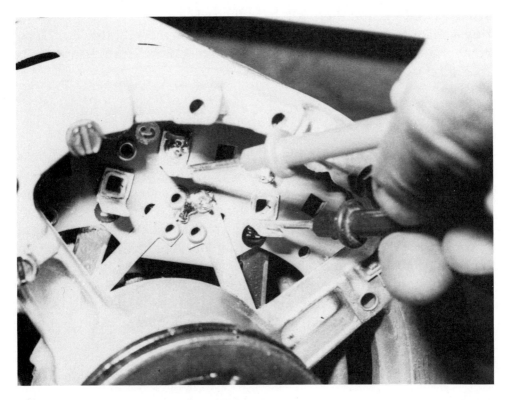

FIG. 5-10 Testing a motor with a VOM.

ing of close to zero ohms), there should be a reading of infinite ohms between different windings. You should also get a reading of infinity between any of the leads and the motor casing. A reading of any resistance here (that is, any reading less than infinite ohms) means that the motor windings are shorting, if only slightly, with the case. This is a dangerous situation.

Since this test is both for the motor and for your own safety, carry it out carefully. Set your meter in the highest range for resistance. In the x1 range, a full-scale reading may or may not indicate infinity. In the x1000 or above range, that full scale reading is more likely to mean what it indicates.

Other electrical parts of the motor can be tested with a VOM. If the motor has a capacitor, for example, the VOM can provide a quick, although not entirely reliable, test.

Short out the capacitor first to drain off any charge it might have. If at all possible, disconnect at least one of the two leads. (The order of events is important here: drain the capacitor of charge *before* either disconnecting it or touching it with your fingers.) Now touch the probes of your meter to the capacitor leads. The needle should swing to a low-resistance reading, and then drift back towards infinity. The meter is putting a slight amount of current into the capacitor. As it "fills up," it becomes more and more resistant to accept more. Hence the change in the reading.

MOTOR DISASSEMBLY

The exact steps for disassembly of a motor will depend on the particular design of that motor. The method used by the manufacturer to assemble the unit must be followed precisely, right down to which screws and spacers go where, and in which sequence.

Go very slowly. Take thorough notes. Make sketches where appropriate. Without notes and sketches, you could get inside, fix the problem, and then be unable to successfully reassemble the motor again.

Keep in mind that many of the motors used today are not meant to be opened. This design is meant to help keep dust and other contaminants out. It also means that such a motor is often impossible to repair.

REPLACING A MOTOR

There are times when an exact replacement motor is not available. If you can find a motor of the same physical size, with a shaft of the same thickness and length, operating on the same voltage, and that you can re-mount in the same position as the original, and in the same manner — then it may be possible to substitute this for the original. However, unless you are well versed in motors and motor types, you are better off working with exact replacements.

In replacing a dc motor, correct polarity (plus to plus, ground to ground) is important. Make sure you solder or attach the screw tap connections to the replacement exactly as they were on the original. Again, your notes and sketches are important. Wire up the new motor exactly the same way as the old one you took out.

MAINTENANCE

There are three basic parts to proper motor maintenance. The first, and easiest, is keeping it clean. The second is keeping it properly lubricated. The third is keeping an eye on all related parts.

Dust, dirt, lint, and so forth are deadly enemies of anything mechanical. Allow a motor to become dirty and it will almost certainly begin to have troubles. Thoroughly clean all motors and related parts on a regular basis. How often depends on how and where the motor is used. A visual inspection every few months is a good idea in any case.

It should be quickly apparent if the suspected motor is designed to accept oiling. These motors need lubrication, but it can present problems if handled improperly. All motors and drive gears should be lubricated at least once every six months (more frequently if suggested by the operating manual). But lubricant must be applied *carefully*. Excess oil can not only damage parts, but it will also collect and gather any dust or dirt in the area, aggravating an already difficult problem.

Use light machine oil (3-in-1 is a good type). Be sure to wipe up any excess spills with a soft cotton cloth. It helps to lightly moisten the cloth with denatured alcohol (*not* rubbing alcohol).

If you get careless and accidentally spill some oil, clean it away immediately. In cases where belt drives begin to slip after an oiling, the inner surface of the belt (where it runs across a drive shaft or pulley) can be "dressed" with either soap or beeswax, just as an automobile drive belt can be dressed when it slips. Just hold the bar of soap or beeswax against the inner surface of the belt as it turns. Excess oil will be absorbed and the belt will begin to "grab" better. You can actually hear the belt-driven device increase in its rotating speed.

In such dressing operations, be sure to use common sense safety precautions. Make sure your fingers, hands, and the soap or beeswax won't get caught in any moving mechanism. Apply the dressing cautiously. Also, check to see if the appliance comes with some type of belt tightening adjustment screws or springs. These may be repositioned to tighten a belt after prolonged use.

If a drive belt still slips after dressing and tightening attempts, it has stretched out of shape and must be replaced. An exact replacement is necessary. Any substitute is only a temporary solution to make an appliance work again. It might *seem* to work okay, but the improper replacement can put a severe torque strain on the drive motor, leading to permanent damage to the motor and to other parts.

This is especially critical in heavy-duty appliances: Replacement motors are a heck of a lot more expensive than replacement drive belts. Don't try to save pennies by using a substitute belt replacement — which will quickly burn out the motor after *seeming* to effect a satisfactory "repair."

Motors driven by ac power often have condensers (capacitors) directly mounted somewhere on the motor casing. These are usually of the *electrolytic* type. This means the condenser is manufactured with an electrical insulating paste between the metal plates of the device. After two or three years' time, the paste begins to dry out and the capacitor begins to leak (allow the passage of dc voltage). A capacitor is designed to *pass* ac voltages and *block* dc. As normal maintenance, electrolytic condensers should be replaced after they are three or four years old, even if they still seem to check okay on a VOM test. This replacement of an inexpensive part can save big costs if the capacitor goes completely bad (shorts out) and destroys the motor or the unit's power supply.

Chapter 6
Heating Elements and Thermostats

All types of ovens, dishwashers, clothes washers and dryers, and electric furnaces make use of some form of heating element. And this is only a partial list. Basically, any appliance that has heat as a function has a heating element to provide it.

Most of the units with heaters are thermostatically controlled, meaning that a device causes an electric switch to turn on or off when certain operating temperatures are reached. Many also have some form of temperature control, usually via something called a rheostat.

The purpose of this chapter is (1) to demonstrate the way these related units work and (2) to show you how faulty devices may be checked and then repaired or replaced.

HOW IT ALL WORKS, BASICALLY

Rub your hands together quickly. The friction will create heat. The more friction there is, the more heat there will be. The less friction, the less heat.

The same happens in an electrical conductor. A wire with low resistance, such as those in the walls of your home, allows electricity to pass through without a large heating effect. Any electrical conductor that offers a high resistance to the passage of electricity becomes a heating element. The energy required to force the current through the resistive wire is passed off into the air around the material as heat.

Heating coils and elements are made of metal alloys of chromium, nickel, tungsten, aluminum, and other metals in combinations that are physically strong and durable as well as heat generating. Nichrome, an alloy of nickel and chromium, is one of the most common alloys used for heating elements.

Various control devices, such as thermostats and rheostats, can be placed in line with the heating element. A thermostat is an automatic switch. When the current flow causes the heating element to reach a certain temperature, the thermostat causes the current to be cut off. With an adjustable thermostat, you can set the thermostat to shut off current flow at whatever temperature you select.

A rheostat can also be put in line to govern temperature. The rheostat is a variable resistor that controls the amount of current flowing through the heating element. The more current, the more heat; the less current, the less heat.

However simple or complex the overall system, there are three basic ways in which heating elements and related parts are internally connected to the unit's voltage supply wiring: plug-in, lug connections, and welded or soldered.

PLUG-IN UNITS

Plug-in units have either prongs similar to those on an electrical cord or specially sized prongs or sleeves shaped to slide over the contact mounts on the appliance. This is the easiest type of heating element to disassemble, test, or replace since removal and reinsertion is a simple matter of unplugging and plugging.

There can be two problems, however. If an appliance has been in use for a long period, the contacts may be almost frozen into place by corrosion, by being slightly bent, or by a build-up of grime that has packed the contact slots. This is especially common when there is water around the heater element and/or the contacts. You'll have to be careful

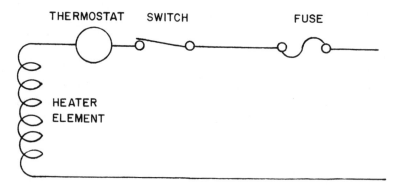

FIG. 6-1   A heating element schematic.

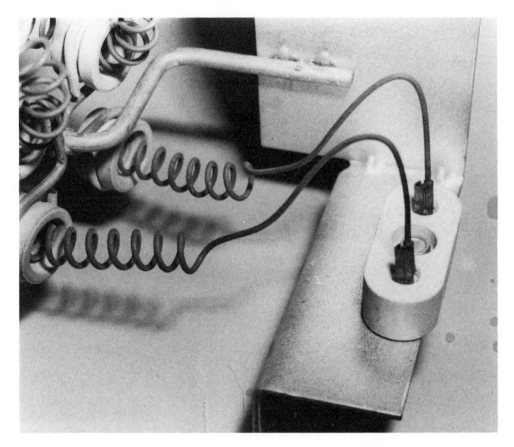

FIG. 6-2 A plug-in heater element.

when removing the plug on such a unit, especially since corrosion might have weakened the prongs.

Don't just assume that the element won't come out because of the build-up. Before attempting to use force to jiggle or pry tight plug-in units from their sockets, study the device carefully to see if the manufacturer has placed small holding or set screws somewhere as a locking device.

LUG CONNECTIONS

Sometimes the wires leading to and from the thermostat or heating element are equipped with spade or eyehole lugs, designed to slip over or under contact lugs, screws, or bolts. That's a complicated way to say that the element plugs in, but with wires and connectors instead of more directly.

Before running out to buy a replacement element of this type, be

FIG. 6-3 A heater element with lug connectors.

sure to inspect the wires and connectors, and then use a VOM to test that the lugs are properly soldered or clamped onto the wire endings, and are making proper contact. Once again, the biggest enemies are grime and corrosion. What appears to be a fault with the heating element could be nothing more than a lug and wire that is no longer making contact. Often an open or partial break in the lug contact or in the lead-in wire itself can be at fault, rather than a damaged or open element.

Wire and lug troubles can be easily repaired at home, avoiding the cost of purchasing a replacement element. You might have to replace the lug if it is too far gone, but the 5¢ cost of a replacement lug is better than spending $15 for a new heater element.

WELDED OR SOLDERED UNITS

Some elements and their connecting wires are spot welded or silver soldered onto the proper voltage contacts within the appliance. This is a particularly common method of attaching the element wire to the overall heating element unit. Replacing such elements and other spot-welded parts can be difficult, and in some cases will be impossible without the proper equipment.

If regular electrical solder is used to install a replacement, the heat generated in making the element work will be sufficient to melt the sol-

FIG. 6-4 Spot-welding on a heater element.

der. To prevent the solder from melting, manufacturers generally either spot weld or silver solder any direct connectors to the coil.

Spot welding is just what the name implies. It is a weld (a joining of metals at very high temperature) in a small spot. The result looks like a small indent or burned spot. This creates an almost permanent joint, and one that won't melt or let go unless the temperature climbs far above the operational level, in which case the element wire will melt long before the weld does.

Silver soldering is a special technique for making heat-resistant electrical contacts almost as strong as spot welding. A solder containing silver is used, which melts only at a very high temperature, and once it is in place, the temperature needed to melt it again is even higher. Once again, before the solder itself melts, the element wire will melt.

To replace spot-welded elements, it is sometimes necessary to use special crimping tools, or heat-generating crimp/clamps. Even if you have the proper equipment, spot welding or silver soldering can be difficult. It depends on the circumstances and where the weld or solder is.

WHAT GOES WRONG?

There's not much to go wrong with the heating element. It's one of those situations where it either works or it doesn't. Diagnosis and testing of the heating element and the appliance as a whole relies on the process of elimination.

Basically an appliance that uses heat consists of a power cord, a switch, a pilot lamp, a thermostat, a rheostat, and the heating element itself. (Not all heating appliances will have all these parts. Some will have other parts and control circuits, which can be diagnosed separately.) Keep this basic plan in mind when you are troubleshooting a malfunction.

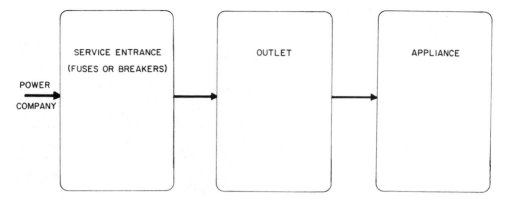

FIG. 6-5 Malfunctions originate in one of three locales: between the service entrance and the outlet; between the outlet and the appliance; or inside the appliance.

For a totally nonfunctioning appliance, or one that is blowing fuses, there are two possibilities. The fault could be between the service entrance and the outlet, or it could be from the outlet and into the appliance. This sounds (and is) so simple that many people ignore the obvious, and will tear apart the appliance, then spend the money to replace the heater element, only to find that the whole problem in the first place was nothing more than a circuit overload.

Particularly with appliances that heat, this first step in the process of elimination cannot be ignored. Due to the heavy current draw of anything that purposely generates heat from electricity, problems of circuit overload are more common. (You can usually skip this step if the appliance has a function other than heating and those other functions work, or if other appliances operate perfectly on that outlet or circuit.)

Once you know for sure that the problem is not in the circuit or outlet, you can begin eliminating the other possibilities. If it's not the cord going to the appliance, then it's *in* the appliance. If the heating element tests okay, then the problem is inside the appliance between the heating element and where power enters the appliance. And so on.

Much of the time the source of the problem can be spotted visually.

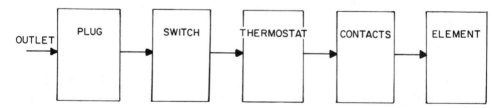

FIG. 6-6 If the problem is inside the appliance, it can only be in a few places: the switch, the thermostat (or rheostat), the contacts, or the heating element.

The heating appliance draws fairly heavy current, and this means that burns, melts, and other damage can occur. For example, if the wire of the heating element is exposed to view, you might be able to quickly spot a melt in the wire. Or you might see a burned contact or other component that has been charred or damaged from heat. (However, wire elements in appliances that create a lot of heat will become discolored and be perfectly fine.)

Probably the most common failure is a burn-out of the element itself. The heater element is made of high-resistance material and is designed to heat up. However, no matter how well made, each time the element heats, stress is placed on the material. Sooner or later, even the best heater element will give out. How it does so, and how soon, depends on the appliance. The element wire itself might melt. Or sufficient corrosion might develop along the element or at the connectors that the current flow needed to create heat decreases or ceases altogether.

The thermostat might freeze up, either on or off. The rheostat might become so corroded that the corrosion forms a layer of insulation between the internal contacts. Even more common, the contacts between the parts, which are often exposed, will corrode and then either break or simply stop passing current.

Fortunately, almost all of these things are easy to diagnose. Repair usually requires a replacement of that part, but that is almost always far less expensive than replacing the appliance itself.

TESTING THE HEATING ELEMENT

The heating coils and their contacts are often visible when an appliance has been taken apart. If the element is damaged from overheating or age, this can often be seen in the form of an obviously melted wire, a crumbling wire and/or insulation, or other such damage.

If you don't see damage, or if the element itself is not visible, make a continuity check with your VOM. You should get a very low resistance reading (although not one of zero ohms) from one lead of the heating element to the second lead. If you get a reading of zero, you're probably not using the VOM correctly. A reading of infinity means that the wire of the element has broken, has melted, or that corrosion has reached a point either in the element or in the contacts so that current flow is seriously impeded.

Most coils resist ac current. This form of electrical resistance, called *impedance*, cannot be measured with the VOM. But if the meter shows continuity from one lead to another, the element is not open, and is

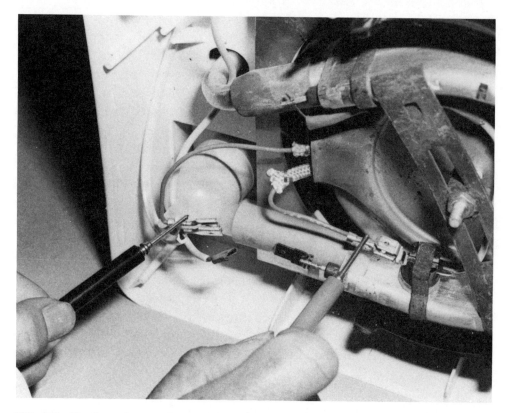

FIG. 6-7 Testing a heater element with a VOM.

probably okay. Check elsewhere in the appliance for that particular problem.

Many heaters and heavier-duty appliances have more than one heating element. In these cases, separate heating coils are used for various temperature ranges, with switching (either manual or via relays) to shift between the various ranges. This is especially true in the case of heavy-duty electric irons and baseboard or wall space heaters.

Other appliances may have two completely separate elements. An automatic coffee maker is a good example, with one heating element used to heat the water, and a second element used to keep the coffee warm.

If the malfunctioning appliance has two elements and both go out simultaneously, it's rather unlikely that both would have burned out at the same instant. Instead of wasting time testing both elements, look elsewhere first. Both elements are supplied from the same outlet. Except for large appliances that require 220 volts, both are probably supplied from the same wires and the same switch. Start your testing with these common items. (Chapter 2 gives you all the information you need on testing cords, switches, and so on with your VOM.)

Before putting away your meter, also test for continuity between both element contacts and ground. You should get no reading (or, rather, a reading of infinite ohms). This shows that the element is isolated from the metal chassis. A reading means that something has shorted out. It could be something as simple as moisture, or something as deadly as a wire touching the metal chassis of the appliance. Either way, if you get a reading between the heater element and ground, *do not* attempt to use the appliance until the source of the problem has been found and repaired.

A problem that is a little more difficult to track down is an appliance that works, but only partially. It might heat up when it is turned on, but not enough heat energy develops to do the job properly. In these instances, only one of the several heating coils or elements may be faulty. Even more likely, the element is simply wearing out. This will show by a high, but not infinite, resistance across the element (as shown by testing with a VOM). Pinpoint and repair or replace the bad element, and the device is at like-new efficiency once more.

REPAIRING THE ELEMENT

When a heater element is malfunctioning, the usual repair is to replace that element. Sometimes this is simply a matter of unplugging the old element and plugging in the new one. Other times it might require removing the element wire and then carefully fishing the replacement wire through the insulators.

Some very simple heating coils, such as may be found in inexpensive room heaters or in small one- or two-burner electric hot plates, may burn open from aging or through being used for too long a time in the high-temperature thermostat range. In emergencies, or when you expect to use the device mostly in the low or mid-range heating mode, a quick repair can be made by slightly stretching the coil so the two open ends slightly overlap and touch each other. Special clamps can be used to hold the two ends together. Such a "repair" will eliminate one or two turns of the coil, so its overall resistance (impedance) will be slightly lower (resistance is determined in part by the total length of a conducting material). The patched coil will not get as hot as a new replacement, but on low or medium heat it may last just as long as the original did on high heat.

An inadvisable method of repair is to pseudo-spot weld the wires.

This involves bringing the two wire ends close together, so that they just barely touch. Carefully apply electrical current, watching the spliced coil. A flow of current will jump (arc) from one broken wire to the end of the other, and this quick arc will spot weld the splice.

THERMOSTATS

The "brain" of any heating device is the automatic control that varies the temperature of the appliance according either to diurnal (day to day) conditions or the need of the operator. This is the thermostat. Several types are in use in any modern household, from the basic heating-cooling thermostat used to control the home's central heating and cooling units, to the small internal devices used to automatically maintain the proper temperatures in hot water heaters, coffee makers, or baking ovens.

There are several types of thermostats, but most work on the basic principle that metals of two different types will expand or contract unequally when heated. Two metals in what is called a *bimetal strip* will bend or change shape according to temperature changes, and this movement is used to operate an automatic electrical contact switch. The opening and closing of this switch turns the heating (or cooling) element on and off.

When bimetallic thermostats are controlled by a rheostat, the turning of the knob changes the relative position of the two pieces of metal. It takes more or less air temperature change to make the two metals expand/contract enough to make the electrical contact.

Very simple thermostats are used as the heat controls for water heat-

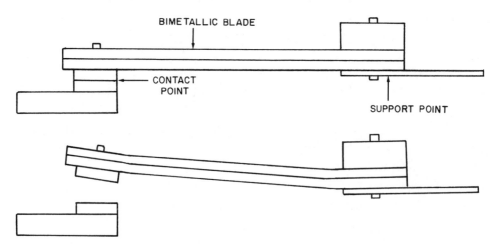

FIG. 6-8 The function of a bimetallic thermostat, closed (*top*) and open.

ers and many coffee makers. In appearance, these seem to be dull silver-colored "lumps" on a short piece of wire. Inside the "lump" is a bimetallic switch. Other times the thermostat will be more exposed; in some the bimetal strip will be quite visible.

A few of the control devices are *thermisters*, or *varistors*, which are semiconductors (transistor-like devices) that are subject to changes in internal resistance when different voltages or amounts of heat are applied. The changing resistance becomes the controlling effect, or "brain" for the appliance.

In total home heating and cooling operations, it is often discovered that a single centrally located thermostat does not do the job of control properly. The spot selected for the thermostat location is too hot or too cool in comparison with the rest of the home. For this reason, many people decide to switch over to multi-stage thermostats, with separate thermostat control used to set temperature for different areas in the home. (Multiple thermostats can be complicated to test, but it can be done by following the process of elimination.)

SOLID-STATE THERMOSTAT CONTROLS

Newer appliances utilize computer-type switching devices. A heat-sensing thermostat can be built right into a tiny IC (integrated circuit) chip, along with transistorized switching networks. As the ambient temperature of the air or the material being heated or cooled changes, the miniature thermostat "reads" the change and "instructs" the switching network to turn the appliance on or off.

These sensing chips are relatively inexpensive and cannot be repaired (although they do go bad). If this is the type of thermostat control on some of your home appliances, you will have to obtain an exact replacement of the chip from the manufacturer or service dealer. And more likely, you will be required to purchase the chip and all of its associated circuitry on its mounting board.

TESTING THE THERMOSTAT

Before wasting a lot of time in testing a suspected thermostat, first visually examine the thermostat and any contacts. The whole problem could be something as simple as dirty or pitted contacts. After you've determined that the thermostat *is* the most likely cause of the trouble, you can go on with testing of the unit.

Since thermostats are switches, and normally open when disconnected, test meters are of limited use in checking to see if the device is faulty. The best test is operational. Once a heating element and other components have been tested okay, and if you know that voltage is being supplied to a device, poor heat control is a sign of a bad thermostat. Repair is usually by replacement of the thermostat unit as a whole. Replacements are usually quite inexpensive—from $3 to $10 in many cases.

Sometimes you can go farther with testing. Other times you cannot. It depends on how the thermostat functions in that particular appliance.

First place a wire across the two prongs of the plug to the appliance (after it has been unplugged from the outlet, of course). This shorts out the thermostat and the appliance in general.

Set your meter to read resistance in the x1 range. Now touch the two leads of the thermostat. With the on/off switch of the appliance in the on position, you should get a reading of continuity across the thermostat. This shows that the contacts of the thermostat are closed, and that current should be flowing.

However, this test doesn't tell you if the thermostat is fully functional—only that it should be allowing current to flow when the appliance is first turned on. If the contacts are fused together, the thermostat will not be able to do its job of shutting off the current. If the thermostat breaks

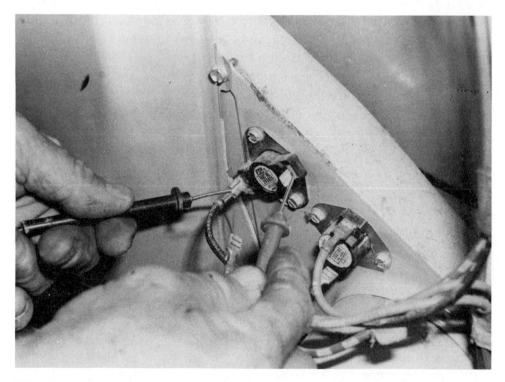

FIG. 6-9 Testing a thermostat with a VOM.

contact too soon, due to maladjustment or wear, the heating element won't get enough current.

Sometimes a meter may be used more successfully for thermostat checking in some appliances, such as a hot water heater. Set the VOM in a high ac voltage range (above 250 vac), and place it across the voltage source (input) to the water heater. Rotate the temperature control rheostat from the lowest to highest temperature setting. At some point, you should be able to hear a light "click" as the thermostat contact is made. At this point, the voltmeter should indicate a *drop* in voltage as a load (voltage being applied through the thermostat to the heating element) is applied. If you hear this click and are able to read this voltage drop, but the water heater still does not go on, then the problem is probably in the basic heating element under the tank. Due to the construction of a water heater, it is best to replace the entire water heater when a heating element goes, especially if the unit is more than eight to ten years old.

Similarly, a voltmeter may be used to check thermostat or thermistor operation in a small coffee maker. With the unit disassembled for metering access, measure the ac voltage from one lead of the electric cord across to the *input* side of the thermostat. If you read voltage there, but cannot obtain a voltage reading from the same side of the ac cord to the *output* side of the thermostat device at a time when that control should

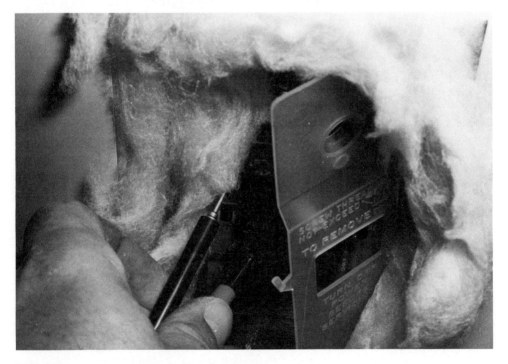

FIG. 6-10   Testing a water heater thermostat.

be turning on a heating element, then the thermostat is likely bad, or the heater element has opened. Repair of either is a matter of replacement.

TIMERS

Although a timing device isn't exactly a thermostat, it can be used to start or stop current flow at preset times. Timers are simple clock mechanisms, with the clock timer on a rotating wheel. Electrical or switching contacts are placed on the moving timer, so that at certain preselected times a heating, cooling, or cooking device will be automatically turned on or off. Most washing machines, dryers, dishwashers and electric ovens have such timers built-in as a basic part of the appliance.

Often, a timer-controlled appliance will have a built-in outlet, much like a wall outlet socket. This means that the unit is designed so the owner may operate a separate device off of that timer, or that the outlet has been designed as a separate voltage source to *bypass* the timer. Examples of this are video tape recorders that may be plugged into a wall outlet. The video machine operates off of an internal timer, preset by the owner. The unit also contains an outlet plug so the TV set can be connected directly to the VCR, but the TV set is *not* on the timer. This outlet serves as nothing more than an extension cord or extension plug for the owner, saving the wall outlet for use by some other appliance such as a lamp.

FIG. 6-11 A switched outlet is used in some appliances to supply power to another piece of equipment.

OTHER CONTROLS

Other controlling switches may be located in conjunction with thermostats. These may be *humidistat* controls to adjust operation according to humidity, or they may be some form of electronic air-cleaning or air-filtering controls. All are switches of some type, and they may be checked and repaired or replaced.

Occasionally, in less-expensive small appliances, temperature is varied either by a switching arrangement or by a variable resistor (rheostat). In switch set-ups, different positions are used to either change the amount of current fed to a particular heating element, or are used to switch the voltage/current feed from one heating element to another.

In rheostat arrangements, a relatively high-wattage (high power dissipation) variable resistor is used as a *voltage divider*. Changing the rotary position of the variable resistor raises or lowers the amount of voltage pressure applied to the heating element. Less voltage means less power will be dissipated as heat energy when the current passes through the heating element. Changing the rheostat to increase the voltage will result in a heavier power dissipation through the element and more heat energy.

MAINTENANCE

Because there are so many switches and moving or slide contacts, poor operation of a heating element can be a simple matter of lousy electrical contact. There are several methods for burnishing, or improving the continuity of a movable electrical contact. Both surfaces may be rubbed lightly with the rubber eraser at the end of a pencil. Hardware stores carry small contact burnishing tools, a type of small file or sanding device that can be used to scrape away grime and oxidation from contact points.

An aerosol spray contact cleaner provides another possible cure. A warning, however. Be *sure* to purchase *nonlubricating* contact cleaner. Several aerosols are on the market that are specifically designed for use on television tuners. They contain a lubricant, or a silicone concentrate, along with the contact cleaner (alcohol or some other fast-evaporating cleansing agent). The lubricated sprays are fine for TV tuners, but if they are used on switch or rheostat contact points in an appliance designed to heat, the lubricant will remain as a dust-gathering contaminant. A buildup of dirt will quickly cause the switch, contact, or rheostat to go bad again, possibly damage it severely.

Most heating elements have no moving parts unless a fan driven by an electric motor is used to disperse the heat. Maintenance is a matter of simple cleaning and keeping dust, oil and grime off the element. Generally no more than once a year, use the reverse blowing setup on a vacuum cleaner or an air compressor to blow dust off these elements, or wipe gently with a dry cloth (moisture is a conductor, and if the coils are wet when electricity is applied, a short may develop and destroy the element).

Simple cleaning of heater elements is a procedure many people tend to forget. Check to determine if the appliance you're working on has an air filter of some type. Filters are employed in forced air systems to clean the air that enters the appliance. While this protects the inner components, those filters can cause a problem. If clogged, the internal temperature can rise, causing the thermostat and other controls to operate falsely. The result is improper operation and possible damage to the appliance.

Filters of the throwaway type need replacement from time to time (check your owner's manual for frequency of replacement). Other types of filters may be removed and cleaned with cool water.

Grills or registers and thermostat devices themselves also require frequent cleaning, since dust and small objects can easily fall into them. Cleaning can usually be accomplished by using a vacuum cleaner or an air blower.

In any heating or cooling device, keep the fan blades clean. When they become loaded with dust and lint, as can often occur if left unattended, their ability to move air is cut down. Most blowers/fans have blades that are easily accessible for cleaning. You can reach them with a small brush or soft cloth. If the blades are not so easily reached, you should disassemble, clean, and adjust the blower/fan at least once a year.

Any belt that powers a fan blade should not be stiff or too loose as it winds around any drive pulley. Properly adjusted, a drive belt should have from ½ to ¾ inch of slack, or play. If the belt appears worn, replace it. A maladjusted drive belt will not only cause a mechanical malfunction, but it could also cause the thermostat, other controls, and the heating element itself, to malfunction.

Thermostats sense the ambient temperature and then turn its associated appliance either on or off accordingly. A coating of dust or grime will make it unable to detect any but the widest temperature changes, and the thermostat will lose efficiency. Clean them annually.

Chapter 7
Appliance Purchase and Care

Most modern appliances require little or no maintenance. With some, the disassembly required for internal maintenance isn't worth the risks.

However, there are some things that you *can* do to lengthen the life of your appliance, while reducing the number of times that you have to call in the serviceperson or replace the appliance.

Most of it is common sense, and proper cleanliness. It doesn't require a lot of time or effort. And for that slight investment, you can increase the life of your appliances while cutting the costs to keep your appliances running perfectly. There is no better way to repair an appliance than to keep it from breaking down in the first place.

Start thinking of preventive maintenance even as you are making the purchase. Will the appliance do the job you want it to do in your home? Does it have the features you need? Will it fit? Is it compatible with the wiring in your home? This involves some careful shopping on your part, and even a few measurements and calculations.

WILL IT DO THE JOB?

It's common for a new purchaser to expect an appliance to do something it can't. This both causes disappointment for the owner, and could tempt the owner to use the appliance in a situation where it could be damaged.

This is the first problem. The appliance you buy should be capable of doing what you want it to do. And you shouldn't try to get it to do things it isn't designed to handle.

The more complex the appliance, the more likely it is that something will go wrong sooner or later. So don't necessarily buy the most feature-filled appliance you can find. If you won't need those features, they are just more things to cause you trouble.

PHYSICAL SIZE

Physical installation is not an issue with most small appliances, but is virtually always something to keep in mind with a large appliance purchase. Will that new refrigerator fit into the spot and still have enough room for adequate ventilation? (Will it fit at all?)

Another related concern is ease of servicing. This is obviously critical with large or built-in appliances. If you can't get at the appliance, repairs will be much more difficult, and the chances of damaging the appliance, your home, or even the repair person increases.

POWER REQUIREMENTS

Be sure you know the power requirements for the appliance you intend to purchase. In particular, appliances that heat should be checked for the amount of current that they draw. The larger the appliance, the more important this is. In some cases you'll have to provide a separate outlet just for that appliance. Power consumption should be listed clearly on the appliance label. If it's not, ask the dealer to show you where the rating is listed in the instruction manual.

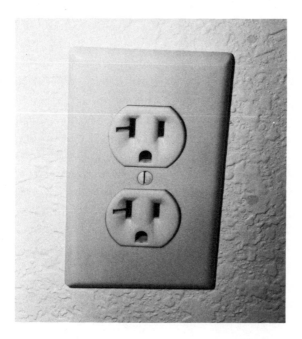

FIG. 7-1 A dedicated outlet. The notch in the outlet indicates that the outlet is for 20-amp service.

The amount of current can be roughly estimated by using the formula $I = P/E$, where I is the current in amps, P is the number of watts consumed by the appliance or other device, and E is the home voltage. (This last will be either 117 or 234.) To use this forumla you'll still need to know the number of watts consumed.

For example, the current used by a 60-watt light bulb is 60 watts/117 volts, or .51 amps (approximately ½ amp.) A device that uses up 1400 watts will draw 1400/117, or about 12 amps.

Be sure not to come too close to the 15-amp rating of the standard circuit. If you plug that 1400 watt device into a circuit, don't think that you can safely load up that same circuit with the 355 watts that brings the circuit to exactly 15 amps of current draw. If the line voltage drops temporarily to 110 vac, the current flow with that 1400-watt appliance will increase to 12.7 amps. Everything else on that circuit will also draw more current. So if the circuit is presently carrying 3 amps (such as six lights), and you plug the new 1400-watt device into the same circuit, you're already at 15 amps total. If the line voltage drops, you'll probably blow a fuse or breaker.

Think carefully about where you will be placing the new appliance, especially in regards to power consumption. If the spot you plan to install the appliance has power available only through a circuit that is already at capacity, the new appliance could easily push that circuit beyond its ability to supply power.

BEFORE YOU BUY

"You get what you pay for" is true in many respects. A difference in price of $10 on a $500 appliance doesn't mean much. But if one appliance has a $500 price tag and a similar one is selling for $300, you can bet that there is a reason. (This rule does not apply to items that are on sale.)

Look for good physical construction, whatever the appliance is. Panels should fit well and be fairly solid. Switches should have a good feel to them, and not be sloppy or loose.

Ask the dealer for a demonstration of how you would go about routine cleaning. Don't believe him if he says, "None is ever needed." That appliance might require very little cleaning or other maintenance steps, but all will require some—certainly more than just wiping down the outside case.

Ease of servicing is also important. This shows a good design, and a manufacturer who has bothered to think things out before putting that product on the market.

One clothes washer comes to mind whenever I think of ease of servicing. The safety switch under the lid, which is designed to prevent the washer from operating if the lid is lifted, was of low quality and was easily broken. To replace that switch, it was necessary to pull the washer away from the wall and remove both the rear and top panels, plus some things inside. What should have been a 5-minute job (unplug the old switch, unscrew two holding screws, replace the switch, screw it back in) took more than an hour. (For that matter, the manufacturer should have used a heavier duty switch in that spot.)

A few manufacturers have taken the time to design their appliances intelligently. Top and front panels can be removed, providing access to virtually everything inside, without having to slide the appliance out from the wall. Even if you don't care to service the appliance yourself, this simple idea reduces your repair costs and also the chances that your home will be damaged as that heavy appliance is slid across the floor.

BRAND SHOPPING

The single best way to know which brand and model to get is to have a personal recommendation from a friend who owns one. If it has worked flawlessly for years for your friends, it's a fair bet that the design and construction are sound.

Some people have a brand name preference; others prefer to save a bit and buy a lesser-known brand that comes with a good reputation or recommendation.

You *do* pay extra for the name, but not as much as you might think. There is usually a reason that the well-known brand costs more. Part of the reason is simply that they can charge more due to their reputation. More important to you is that the extra cost pays for a number of services.

The manufacturer might have large research facilities, used to find and correct faults in previous designs. Others have special service schools for their field representatives. A few even have toll-free numbers for their customers to use. More important, and also as a general rule, the major brand names will make it easier to find parts, and to find service when it is needed.

THE DEALER AND THE WARRANTY

When you buy a new appliance, you are also buying a warranty that says that the appliance will work as specified for a certain period of time. If it doesn't, it will be repaired free of charge.

You can run into several problems here. First, if an appliance is still under warranty, *do not* attempt to repair it or even open the case. This will probably void the warranty. It can be expensive to find out.

Before you purchase the new appliance, ask what the warranty period is. The longer this period, the better.

Just as important as the warranty is to know who does the servicing. If the dealer doesn't, he should at least be able to tell you who does. If he has no idea who can handle the servicing—and particularly servicing while under warranty—you might start considering going to another dealer.

It is an unfortunately common practice these days for the dealer to have no servicing available. One of the worst of the new policies you see is the tag on the new appliance, "Do not return this appliance to the dealer. Send it directly to the manufacturer." But that manufacturer might be on the far side of the country, or even outside the country. Shipping costs can get expensive very quickly. And if you don't properly package the appliance, it can get damaged in transit. And it takes a lot of time and trouble to ship things.

Many appliance owners don't even bother. They'll go out and buy another coffee unit—one from another manufacturer, hopefully. If the appliance in question is very expensive, or very large, the problem is compounded. Throwing away a $25 coffee maker is one thing. Tossing out a fancy $400 microwave is quite another.

Consequently, it is usually better to make your purchase from a dealer who also offers warranty repairs and long-term service. However, just because the dealer has a service bay doesn't mean that he will provide warranty work on your purchase. If that small appliance fails, and it says right on the box, "Do not return to dealer," he is not obligated to provide warranty work.

If you are promised warranty work, be sure to get it in writing. Otherwise it could be just a sales ploy, and you'll find yourself shipping off that appliance to a distant manufacturer anyway.

THE INSTRUCTION MANUAL

All new appliances come with a variety of papers, stickers and tags. It's a good idea never to throw out anything that might be of use in the future, especially any instruction or installation manuals. Make sure that you have one. If you get home and find that something is missing, contact the dealer and the specific salesclerk.

Then read that instruction manual! On more complicated appliances, and large appliances, read the installation instructions, even if you didn't personally handle the installation.

Even if you have a separate identification tag, jot down the model and serial numbers on the instruction manual. Sooner or later you'll want these, and it's good to have all the information in one place. Some manuals will have at least the model number already on it; other manuals cover a variety of models. In this case make note of the specific model number. Also put all receipts with the manual.

You're not done yet. Before you put the manual in the file, be sure that anyone else in your house who will be operating the appliance, or who even *might* be, has also read through the information and understands how to operate that appliance.

In some cases a shop manual will be available. The cost for this manual can be steep. You'll have to decide for yourself whether or not you want to spend that money. At very least, inquire about the availability of such a manual. If you don't want it now, you may in the future.

INSTALLATION

Proper installation may not sound like a maintenance step, but it is. It's also for safety. With some appliances, installation involves nothing more than plugging it in. Others require a more complicated installation.

You should have little trouble installing any standard home appliance yourself. By handling the installation yourself, you also know exactly how things went in, and you'll know how to *un*install the appliance (and how to reinstall it) if servicing becomes necessary.

If someone else is going to be handling the installation, try to be there to make sure that it's done right. Whether you're there or not, go through the installation instructions and double check the work. (If *you* did the job, triple check it.)

There are three basic parts to installation: electrical, plumbing, and leveling. Not all appliances require all three.

Electrical installation involves everything from the service entrance to your home to the contacts inside the appliance. (Yes, there *are* times when the manufacturer expects you, or the installer, to open the appliance and make the final connections.) As mentioned earlier, the outlet has to be of the right type, and has to be able to safely supply the appliance. Grounding is very critical. Do not try to defeat the safety grounding by doing something silly like clipping off the third prong. If you don't feel competent to provide the correct grounding, spend a little extra to have a professional electrician come out.

If the appliance uses water, you'll have to bring water to it, and will possibly have to take waste water away. Improper installation can make a mess. It can shorten the life of your appliance by causing rust and corrosion. Worst of all, the combination of water and electricity can be deadly.

Venting involves more than just running pipes outside. It includes providing proper and adequate ventilation for the appliance. Venting is to take away unwanted dust, lint, moisture, or heat.

The installation instructions, and usually the owner's manual, will tell you if special ventilation is needed, and if so, how much. If the appliance has an internal fan, proper venting could be a critical factor. Don't go for the minimum. It's impossible to have ventilation that is too good, but all too easy to have venting that is inadequate. Inadequate venting spells trouble, the least of which is a shortened lifespan of the appliance.

Proper leveling is important for keeping the appliance functioning well. Even if someone else installed the appliance initially, check the leveling. And repeat the check occasionally. It takes just a few seconds. If the appliance is level, fine. You've invested a very short bit of time to know that. If it's unlevel, you could add years of life to that appliance by leveling it.

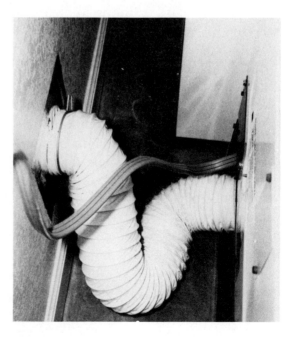

FIG. 7-2 Proper ventilation of large appliances is essential: follow manufacturer's recommendations.

GENERAL MAINTENANCE

Talk to any service technician and you'll hear again and again the value of preventing problems. A few simple maintenance steps might take ten minutes now. Not doing them can cost you hundreds of dollars later.

The first step is cleaning. How you go about this will depend on the appliance and the circumstances. For a coffee maker this will mean regular cleanings with vinegar or a commercial cleaner designed for that purpose. A refrigerator will require a regular cleaning of the coils to keep them free of dust.

Use common sense. As always, your own safety comes first. For example, don't start swabbing down an appliance if it's still plugged in and if the water can get anywhere near a live contact.

Think about what it is you're doing before you just lunge ahead. Some parts are delicate and can be damaged by rough handling.

Be sure to use the appropriate cleanser for the job. A harsh, gritty cleanser is unsuitable for cleaning a shiny, porcelain surface. It can also scrub off many of the nonstick surfaces used in modern appliances.

Obviously you won't be taking your appliance apart every couple of weeks just to clean out the dust. Most of your cleaning will be limited to the exposed parts. But occasionally it's worthwhile to do some interior cleaning on many appliances.

The idea is to reduce the amount of time and work. If something has gone wrong inside and you have to disassemble the appliance anyway, spend a few extra minutes to clean the inside even before you begin working. (It will make working on the appliance easier and safer anyway.)

Any time the appliance needs to be opened, give the insides a checkup. Carefully, with the appliance unplugged, visually and manually check things over. Quite often you'll be able to see something that is starting to go wrong. A switch might show signs of burning; a drive belt might be sagging or show signs of damage; there might be water inside, or excessive corrosion, showing a plumbing problem.

This is a good time to lubricate any motors that can be lubricated. Be very careful to not overlubricate. Usually a drop or two will take care of things. Use a clean rag to wipe away any excess or any spills. Many of the motors used in appliances are sealed units, and generally they require no lubrication.

Whenever you do anything, make notes and sketches for future reference. The notes should include a record of all maintenance done, and anything you've noticed that suggests a possible future problem. For ex-

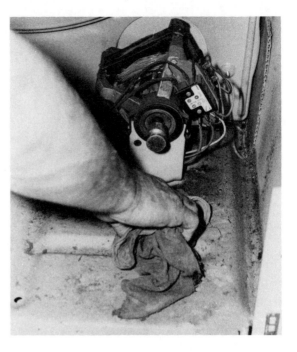

FIG. 7-3 If the appliance is open anyway, take a moment to clean the interior and to check it over for signs of wear that may lead to future trouble.

ample, if a switch shows a bit of discoloration, make a note of that. Next time you look inside see if it has become any darker (or lighter).

DAMAGE TO THE CASE

Most of the physical damage done to appliances is the result of carelessness. Someone drops a hammer and chips the porcelain, or a board is rammed into the side of the appliance. Your primary concern with any major damage to the case isn't cosmetic, but the effects on the inner workings of the appliance. If that dent causes some moving part inside to rub or bind, don't try to operate the appliance until the damage is repaired.

Some case damage *cannot* be safely repaired. You may be able to buy a replacement part; if you can't, and if the damage is affecting the operation of the appliance, it is probably time to go shopping for a new one.

Some case damage *can* be repaired. Chips in porcelain are an example. Chips in porcelain are usually just cosmetic. But they can also lead to rusting of the metal panel beneath and to other problems. Bottles of liquid porcelain are available at most appliance stores and hardware stores. The liquid is simply brushed on.

There are two keys to using liquid porcelain successfully. The first is in the preparation. The liquid will not adhere properly to a surface that is dirty or wet. Second, take your time, and do your best to have the patch job blend in. Blobbing it on makes a noticeable and unsightly repair.

Better than brushing it on by hand is air brushing. Most people don't have the skill or equipment to do this properly, but it is an alternative. If your own attempts aren't working out very well, consult the Yellow Pages in your area under Appliance Refinishing. It's expensive, but it may be less expensive than buying a new appliance. Explain the situation over the phone, get an estimate, and make your decision from that estimate.

Small dents might be repairable. Sometimes you can simply pop a dent out by pushing from the inside. Other times a dent puller (for automobile bodies) can be used. In many cases dents are not repairable, due to the size and nature of the appliance.

Section Two

Appliances

Chapter 8
Refrigerators and Freezers

The refrigerator is a standard item in today's household. Many homes in fact have not only a refrigerator with its built-in freezer compartment but also a separate stand-alone freezer. Some also have smaller refrigerators for the wet bar or out in the motor home. Basic operation, maintenance, troubleshooting, servicing, and repair is the same for all of them.

HOW IT WORKS

A refrigerator or freezer works on the principle that a gas cools as it expands and gives off heat when it is compressed.

A refrigerator or freezer has four basic parts: the compressor, the tubes carrying refrigerant, a fan, and a thermostat.

A compressor is used to compress or "squeeze" the refrigerant and usually also to pump it. A length of tubing is connected to the compressor and provides a completely sealed path. The tubes are filled with the refrigerant, which is usually Freon, either type R-12 or type R-22. This tubing is divided into two basic sections: the evaporator and the condenser. The names tell you just what is going on in that section.

In the evaporator, the liquid Freon vaporizes into gas. It evaporates. In doing so it becomes cold and is capable of absorbing heat from the inside of the refrigerator. Evaporation occurs because the liquefied Freon flowing into the evaporator is flowing into an area of low pressure.

It then passes through the compressor and into the condenser. The compressor squeezes the vaporized Freon and puts it under high pressure. As the pressure increases, the Freon loses the heat it has picked up in the evaporator side. This is helped along by cooling fins or wires attached to the tubing and often by a fan that blows air across the coils. The

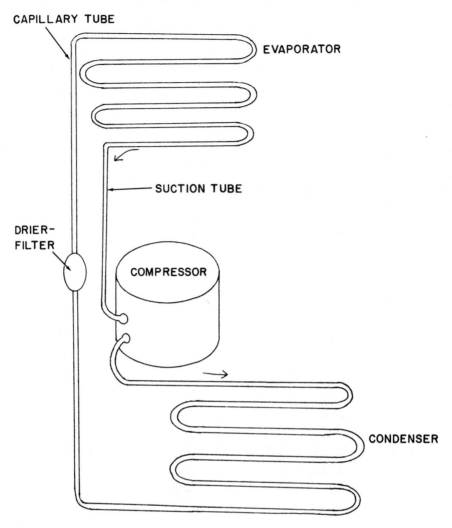

CAPILLARY TUBE

EVAPORATOR

SUCTION TUBE

DRIER-
FILTER

COMPRESSOR

CONDENSER

FIG. 8-1   The function of a refrigeration cycle.

increasing pressure and the lowering temperature turns the vapor back
into a liquid. It condenses.

To further increase this action, a capillary tube is used to connect —
or rather to separate — the high pressure condenser from the low pressure
evaporator. As the refrigerator shuts off from its cooling cycle, you will
hear a soft whooshing sound. This is the rebalancing of pressure between the two components.

A thermostat is used to measure the temperature inside the refrigerator and to trigger the compressor and fan. If the temperature inside exceeds the setting, the compressor comes on and the cooling cycle begins
all over again.

Most refrigerators and freezers also have an automatic overload protection device. For example, if the refrigerant leaks out of the system, or if the coils on the condenser become so clogged that the Freon no longer condenses back into a liquid, the evaporator side will become warm (rather than cold). Then the overload protector shuts down the system — hopefully before damage is done.

Because of this overload protector, if the refrigerator has been turned off for a while, it might take what seems to be an excessive amount of time before the compressor comes on once power is applied again. This is normal.

Simple refrigerators have the evaporator located in the freezer section. Since cold air is heavier than warm air, it drops down from the freezer and into the refrigerator compartment of the unit. More complex units have a split evaporator, with an upper section for the freezer and a lower section for the refrigerator. (These can also be placed side-by-side.

TROUBLESHOOTING

Caution must be exercised when working with any refrigeration appliance. The Freon inside the tubes is under pressure. It can be dangerous under certain conditions and can turn into a poisonous gas. Many of the problems involved require special equipment and background and are best left to a professional.

Whenever a freezer or refrigerator appears to have stopped working, open the door and see if the interior light comes on. If it does not, check to see that the unit is plugged properly into the wall outlet. If the light is working, the appliance is getting power. If the light is not working and neither is the refrigerator, the problem is probably with the power supply (fuses or breakers, outlet, plug, wire connections, etc.).

If the plug is in and the fuse or breaker appears okay, plug another appliance known to be in good working condition into the outlet normally used for the refrigerator. Or test the outlet with your VOM set to read the incoming 120 volts ac. If the outlet is good, but nothing works, find where the power enters the appliance and inspect and test the terminal block and contacts. Since there are so few electrically powered components in the refrigerator, tracing the wiring is usually fairly simple, even without a wiring diagram. You can test the wiring for continuity (with the power off) and test various points (*carefully*) for ac power. If you still haven't found the problem with a completely dead refrigerator, it's time to either call in a professional or to replace the unit.

When there is power coming into the unit but cooling isn't taking place, begin (as always) with the obvious. Take a look at both the fresh

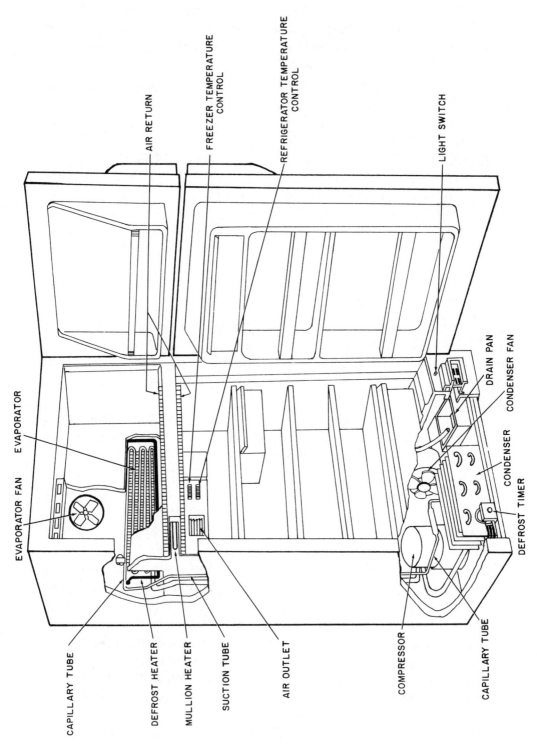

EVAPORATOR FAN   EVAPORATOR

CAPILLARY TUBE

AIR RETURN

FREEZER TEMPERATURE
CONTROL

REFRIGERATOR TEMPERATURE
CONTROL

LIGHT SWITCH

DRAIN PAN

CONDENSER FAN

CONDENSER

DEFROST TIMER

CAPILLARY TUBE

COMPRESSOR

AIR OUTLET

SUCTION TUBE

MULLION HEATER

DEFROST HEATER

CAPILLARY TUBE

FIG. 8-2 Location of the parts of a refrigerator.

food and the freezer temperature control dials, which may have accidentally been turned to lower values, or might even be completely off.

Next examine the condenser coils on the back of the refrigerator. (For a refrigerator that uses a fan, these coils might be hidden beneath. An access panel is generally present at the bottom and in the front of the unit. This panel usually clips into place and no tools are needed for removal.) If the coils are dirty, their capacity for dispersing the build-up of heat will be reduced. Not only will the refrigerator be unable to handle its job efficiently, you could actually be causing severe and expensive damage.

Clean the coils on a regular basis. Use caution when doing the cleaning, however, since the coils and cooling fins or wires are delicate and can be damaged. Once they are damaged, there isn't much that the homeowner can do other than to call in a professional. And the professional might not be able to do anything more than to recommend a new unit.

Check the drip pan. Even though it seems to be external to the system, the drip pan often serves to help direct the air flow. If it's out of place, temperatures inside can rise. (The drip pan can also cause odor in the refrigerator if the pan is allowed to get too dirty.)

Another common cause of a refrigerator that is too warm is frequent door openings. Each time the door is opened, some of the cold air inside will escape. The longer the door is open, the more cold will get out. Excessive moisture or frost collecting inside is also caused by too frequent or too long door openings. Hot and humid weather increases the rate of

FIG. 8-3 The cooling plate inside the refrigerator can be inspected for signs of damage or over-frosting.

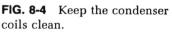

FIG. 8-4 Keep the condenser coils clean.

frost or moisture build-up and internal wall sweating—and this is just when people are most likely to be opening the refrigerator or freezer most often.

Each door or sliding tray is generally fitted with some type of rubber or plastic gasket, helping to ensure a tight fit and cooling economy. Care-

FIG. 8-5 Condenser coils are often located beneath the cabinet.

fully check the gaskets themselves. As with any gasket, a simple cleaning might cure the entire problem. Look through any information pertaining to them in the operator's manual. This will help you to determine the correct way to repair or replace them if they become damaged sometime in the future.

Refrigerators operate with a liquid coolant running through small pipes built into the interior of the walls. Occasionally, this cooling tubing leaks, or refrigerant levels drop from normal operation over time (fortunately a very long time). Testing or refilling the refrigerant usually requires the use of special equipment, and a special valve since the tubing is usually permanently sealed. Checking the refrigerant level not only requires special equipment but it presents certain risks — mostly to the appliance but to yourself as well. It is best to have a professional technician check for leaks and handle proper recharging of a low liquid coolant supply.

A leak may or may not leave visible signs. It doesn't take much of a hole for the refrigerant to leak out. However, the leak often leaves an oily spot behind. (Don't confuse this with a normal build-up of grease that comes from the unit simply being in the kitchen.)

FIG. 8-6 The refrigerator motor and capacitor.

Another clue to low coolant levels is that the refrigerator attempts to cycle but shuts itself off almost immediately. This is brought about by the overload protector. If the refrigerant level is too low, this protector will automatically disable the unit and nothing will cool.

However, the problem could also be in the protector itself. Failure to start the cooling cycle could also be caused by a faulty start-up capacitor, or a shorted motor. You can't easily test the overload protector, but this is rarely the cause of failure. You can check the capacitor and motor (see Chapter 5).

If you haven't found the problem at that point — after eliminating all the obvious things, and after testing the capacitor and motor — call in a serviceperson. It's too easy to cause expensive damage if you proceed without the proper equipment and background.

INSTALLATION

Installing a refrigerator is a little more than just sliding it into place and plugging it in. There are several important considerations, such as adequate ventilation, proper surrounding temperature, and leveling.

Older models use static air flow to carry heat away from the condenser. The coils will be mounted on the back of the refrigerator. This requires a cooling space at the back for proper ventilation, and a space at the top of the refrigerator. The owner's manual will specify how much space is required in both places. Don't make the mistake of putting so much on top of the refrigerator that the air flow is blocked.

Newer models have a forced air system, with a fan to provide air flow over the coils. Most of the venting takes place along the bottom of the appliance. (You can feel the warm air moving along the floor when the refrigerator is operating.) Such units don't have coils exposed on the back of the appliance. The refrigerator may be placed flush against a wall and requires no more than a fraction of an inch of clearance space on the sides.

Some people make the mistake of thinking that if the unit works well in a warm kitchen, it will work even better out on a cool patio. Most units are not designed to operate under cold temperatures (less than from 40 to 60 F). And during the summer, it may be too hot outside for the unit to operate. For this reason, determine the operational characteristics from the dealer before deciding to use outside the pool or patio. Some refrigerators will not work well outside.

There are leveling screws at the corners of most models. Sometimes only the front legs can be moved, which could mean that metal or wood shims will be needed for the back legs. Use a spirit level or carpenter's level to insure that the appliance is firmly in place. It should be level from side to side and with the front just slightly higher than the back so that the door closes gently on its own. (Most leveling screws turn counterclockwise to raise, and clockwise to lower.)

Proper leveling will eliminate excessive vibration noise and improve the operational efficiency. This is especially true with units that have extras such as a self-filling automatic ice tray or cold water or beverage taps. Also, the doors may be difficult to open or close if the appliance is not level.

Both refrigerators and freezers may be purchased with doors that open either from the left or from the right side. A few models have doors that can be switched, either by the dealer or at home following the instruction booklet. Some of the newest models of refrigerators and freezers are constructed with double-hung magnetic doors, which may be opened either left hand or right hand at any time, with no changeovers necessary at all.

Refrigerators and separate freezer units should always be plugged into an individual electrical wall outlet (120 volt, 60 cycle ac). It is best that no other kitchen appliance or lamp be operated from the same outlet.

When first installing and turning on a new refrigerator or freezer, there may be several control functions to set. Most units have separate temperature controls of the rheostat type for the fresh food areas of the unit and for the freezer cabinet. Also, some models have individual temperature selector dials or switches for meat trays, vegetable crispers, butter trays, and for any built-in water coolers or cold beverage dispensers. Most of the control knobs have an "average" marked on them. This is a good place to start. Later, if the temperature inside is too cold or too warm, slight adjustments can be made until the temperature is just right.

If the unit has a self-filling automatic ice maker, the cold water tap of the appliance must be connected to the household plumbing, usually by using ¼-inch copper line, and the separate water switch for the ice maker or cold water dispenser must be turned on. (If your unit *does* have a water switch, it must be turned off any time that the water supply is disconnected or turned off. This is because the water valve could be permanently damaged if the ice maker tray's lever accidentally drops down and "calls" for the ice trays to be refilled.)

Study the owner's manual for your particular model to determine if there are any special venting controls or louvers that may require adjustments for various appliance functions.

ABANDONED REFRIGERATOR

An uninstalled and unused refrigerator can be one of the most dangerous appliances in the home. If an old refrigerator is not in use, be sure to remove the doors! This is to reduce the possibility of danger to small children. Most recent refrigerator/freezer models have specially designed doors that help to prevent children being accidentally locked inside.

For normal vacation periods of up to four weeks, no special attention need be paid to the modern refrigerator or freezer. Under frost-free automatic operation, they can be left unattended with controls at their normal settings. If you are to be away with the appliance unused for periods of over a month, then remove food, turn all temperature controls off, clean and wipe the interior dry, and leave the doors open in order to prevent odors.

If you are moving a freezer or refrigerator from one location to another, first remove all stored foods and thoroughly clean the appliance. Then remove all shelves, trays, drawers, the evaporator pan, and any other loose parts and pack these separately. Tightly tape any moving parts such as the lever for the automatic ice maker into position. Disconnect the power plug from the wall outlet, and always be sure to handle the appliance with care.

CLEANING

Most refrigerator/freezer shelves and trays lift or snap out easily, so that they may be cleaned in the kitchen sink right along with dishes. Use detergent and a nonabrasive scouring pad on stubborn stains. The exterior of the appliance may be wiped with a damp cloth, then dried. Occasionally you may wish to use one of the special appliance enamel waxing or cleaning compounds which will help to renew the finish. Never use harsh or abrasive cleaners or pads on the enameled surfaces. In case of accidental scratches, enamel touch up spray is available at most appliance, hardware, or department stores.

In removing shelves and trays or drawers for cleaning, be sure to study the way the devices are to be inserted and removed. If available, consult the owner's manual. Using excessive force to lift or pull out a shelf may result in damage to either the item or the wall bracket. Many of these clips or brackets are plastic and they will break under excessive pressure or force.

In any cleaning, use warm but not hot water, since high temperatures may cause some of the plastic components to warp or discolor. When cleaning the interior, the normal baking soda mixture is about a teaspoon of soda to a quart of water. After cleaning, it is good to leave either a small box of opened baking soda, or a small amount open in a saucer, somewhere in the fresh food compartment. This will help in the elimination of food odors. To assist in curbing these odors and to prevent the drying of stored foods, be sure to use containers with tight-fitting covers for all foods placed inside the unit.

The life of door gaskets may be enhanced if they are periodically cleaned with warm soap and water, and then with the baking soda solution. Rinse with clear water and wipe dry.

DEFROSTING

Long gone are the days when weekly or monthly the refrigerator and freezer sections had to be completely emptied, the unit turned off, and drip pans inserted on the shelves and refrigerator floor in order to catch melting ice from the freezer section. Now, automatic defrosting is standard in even the most inexpensive models of refrigerators or freezers. Preset factory controls permit the units to defrost automatically on a regular basis.

During the defrost cycle, there is normally only a slight change from

FIG. 8-7 Be very careful when removing the drip pan for emptying or cleaning. Shown is the refrigerant capillary tube leading into the drier: both delicate parts are often located close to the drip pan.

the normal sounds coming from the working appliance. The defrost water is channeled down to the base of the unit into an evaporator pan, where fans or blowers automatically evaporate the melted liquid. That old-fashioned and messy defrosting chore is literally a piece of history.

However, it may be necessary to clean the drip water channels and the evaporator pan and grill (at the bottom of the refrigerator) on a regular basis. Just wipe the interior of the unit with a soft cloth moistened with a mixture of baking soda and water and use a vacuum cleaner to pull any dirt or lint from the grill and evaporator tray (the grills usually just snap off in order to obtain cleaning access to the evaporator tray).

Troubleshooting Guide

Possible Causes	Solution
Problem: Unit does not turn on. Interior light does not go on.	
No power to outlet	Check outlet
Faulty power cord or plug	Check cord and plug
Problem: Breaker or fuse blows when unit is plugged in or turned on.	
Circuit overload	Try on another outlet supplied by another circuit; use dedicated outlet
Short circuit	Test appliance and house wiring
Problem: Unit turns on, but does not cool properly, or does not function in accordance with the operating manual.	
Defective relay, thermostat, or overload control	Check; replace
Motor winding or condenser/ compresser element bad	Test; replace
Connecting wires from power supply are open	Test

Chapter 9
Washers and Dryers

To effectively cover both washers and dryers, this chapter is divided into three basic parts. The first covers washers. the second discusses dryers, and the third part covers topics related to installation, moving, and maintenance.

WASHERS

HOW A WASHER WORKS

Most of today's washers, even the least expensive ones, have pre-wash soak and/or rinse cycles, detergent and bleach/dye dispensers, water temperature controls, water level controls for conserving water when washing small or medium loads, lint filters, and a series of washing cycle controls to vary the action of the agitator and the spin cycle.

An automatic washer basically does what you would do if you were going to wash clothes by hand, but it does it faster: clothes go in, water comes in and the clothes are washed, dirty water goes out, clean water comes in, clothes rinse, water goes out, and clothes come out.

When you set the selector switch and apply power, a double solenoid valve is activated. Water is allowed to flow in to the preset level, at which point the solenoids slam shut and cut off the inflow of water. (Many washers also have an overflow sensor or some system to handle an overflow of water if the solenoids fail to shut off the water.)

Next the washing cycle is activated. The agitator moves back and forth, and sometimes up and down as well, to clean the clothes. Agitator action is driven by a motor, a system of gears and/or belts, and a cam to create the reciprocal (back and forth) motion needed. Meanwhile, in

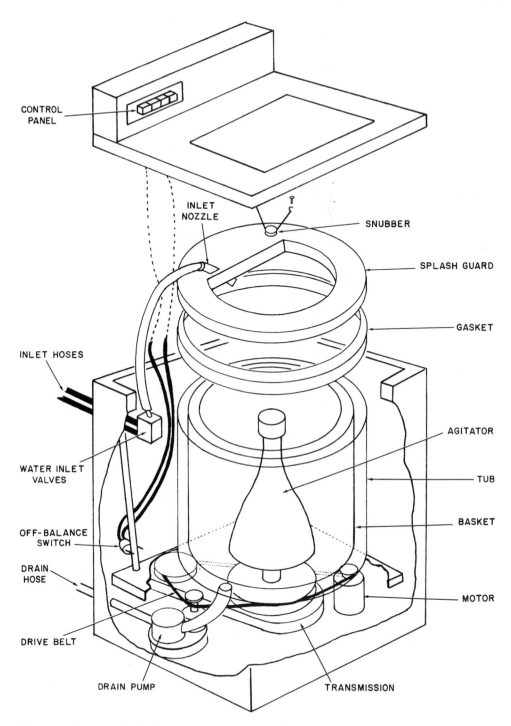

FIG. 9-1 The location of parts in a clothes washer.

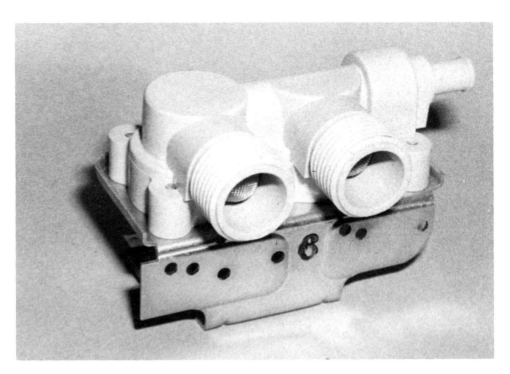

FIG. 9-2 A water valve assembly.

most units a pump is used to recirculate and filter the wash water. (A good way to "inspect" the pump is to look at the recirculating filter. If water is flowing through it, the pump is probably good.)

The pump is activated and the wash water is drained from the tub. To reduce the amount of water even more, the motor causes the tub to spin. The water is pulled out of the clothes by centrifugal force, leaving the clothes damp instead of soaking wet.

The inlet valve solenoids trip again and once more the tub fills with clean water to begin the rinse cycle. (Some machines have only one rinse cycle; others have the option of a second rinse.) Normally there is a soak period and then the agitator finishes the job of rinsing out the remaining soap. The rinse water is then drained from the tub and the clothes are spun once more to remove as much water as possible.

During machine operation, short pauses occur as the washer moves from one cycle to another, or even in the middle of a particular wash or rinse cycle. These pauses are part of normal operation of a servo-mechanism inside the appliance as it changes from one mode to another.

All washing machines have a lid safety switch, which stops the unit automatically if the lid is opened during a spin cycle. In most cases, the safety switch is a mercury switch—a bulb containing liquid mercury that makes or breaks contact as the liquid metal is moved by the lifting of

FIG. 9-3 Location of the safety switch in the washing machine's lid.

the lid. In other cases, the safety switch is a contact interlock that automatically breaks circuit contact (leaves an open circuit to turn washer off). In any case, you should never reach inside a tub to remove, add, or straighten the clothes until the spinning stops. This is for your own safety, and also for the safety of the machine.

In wash/rinse cycles, the agitator will continue to function when a lid is opened. Even so, always stop the washing action before adding any additional clothing.

TROUBLESHOOTING

The "transmission" in an automatic washing machine is a servo-mechanism that can either be a geared transmission similar to that in an automobile or a series of solenoid switches. The transmission shifts the washer from one cycle to another as the selector knob timer rotates from start to wash to rinse to spin dry to stop. It activates the agitator action when called for, puts the agitator in "neutral" for spin dry cycles, and stops tub movement during the water filling periods.

Most washer motors and transmissions are of the lifetime self-lubricating type, and they are sealed, with no visible oil holes. However, washer life can be prolonged and machine efficiency improved if at least twice annually you disconnect the appliance and carefully lubricate the motor shaft and all moving parts in the bottom chassis section of

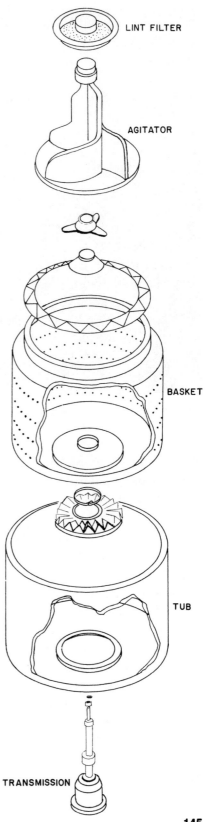

LINT FILTER

AGITATOR

BASKET

TUB

TRANSMISSION

FIG. 9-4 Interior parts of a washing machine.

145

FIG. 9-5 Washer drive mechanism.

the washer. Be sure all water and electrical connections are turned off during this procedure.

At the same time, check all drive belts for wear and tightness, and visually inspect all moving parts to check for wear or for abrasive rubbing and bent or damaged components. Such a check should always be made immediately if any strange whining or thumping noises are heard from the interior of the unit.

For details on motor troubleshooting, refer to Chapter 5. A washer motor has compound windings, and the continuity of each winding must be tested for opens and shorts in case of motor malfunction. In case of an excessive grinding noise, test the motor bearings and see if they can be helped by lubrication or if they must be replaced.

Incomplete draining is most often caused by a clogged or damaged drain line. The small holes in the tub help to keep items of clothing inside, and to prevent them from getting into the drain pipes. But it can happen. More likely, the drain hose is kinked or simply clogged by a build-up of lint. Shut off all water and power and check these things first.

The pump is also suspect any time the washer draining action is slow or incomplete. Once again the most usual cause of a pump malfunction is some kind of blockage. Despite filters and safety devices, a stocking or small piece of fabric may have become caught either in the pump hose, the filter, or inside the pump action.

The pump is another device that should be cleaned and checked at least semiannually. To service the pump, turn off the faucets and disconnect the pump drain hose. Disconnect the pump input hose. Test to see that water can flow freely through both hoses, and then reconnect the hoses and determine if the pump is flushing water out under pressure during drain cycles of the machine. A stiff wire probe may be used to unplug drain hoses. A plugged or malfunctioning pump will have to be disassembled in order to find out exactly what has gone wrong. Sometimes the trouble is a simple, inexpensive gasket or filter component, easily replaced. Other times, the pump or pump hoses will have to be replaced.

As stated earlier, if the water is flowing through the lint filter, the pump is probably okay. Make it a regular practice to look every month or so.

The pump, like the motor and transmission, may have moving parts

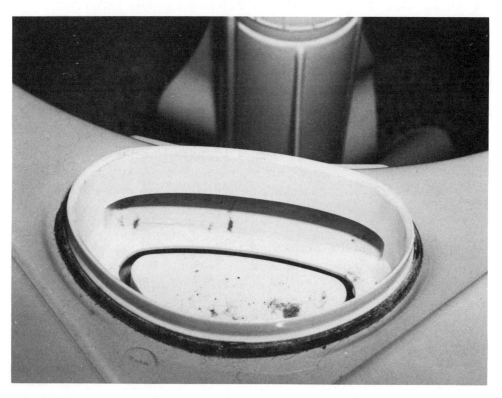

FIG. 9-6 If water is flowing through recirculating filter, pump is probably good.

FIG. 9-7 Washer pump.

that will benefit from occasional lubrication. But be sparing in the use of light machine oil; it can stain enamel surfaces or the floor and get on belts or drive pulleys and lead to slipping in the transmission. Wipe off any lubricant carefully, and be sure to dress the inner surface of all drive belts with soap or beeswax after completing any lubrication.

You may detect a whistling or whining noise when a washing machine is spinning. This sound is normal if the tub is empty or the load is quite light. If the load is heavy, that whining sound can be caused by the belts slipping. Neither is too much of a cause for worry. If the whining continues under all conditions, check the drive belts first, then the transmission, and finally the motor.

If a washer starts vibrating violently, thumping against the floor, it is a sign that the load has become out of balance — too many clothes are on one side of the agitator. In sophisticated models, a safety out-of-balance drive takes over to slow the machine action and a buzzer sounds before the action becomes violent enough to shake the machine or cause damage.

If the buzzer is not noticed, or if the spin cycle is completed, the clothes may still be quite wet. This can seem to be a problem with the pump when it is not. All you have to do is to rearrange the load for balance, turn the selector knob off, select a spin dry cycle, and turn the selector knob on again.

WATER PROBLEMS

Finding and fixing water problems in a clothes washer is almost identical to finding and fixing the same with a dishwasher. As always, begin diagnosis with the simplest things. If the washer fills slowly, the problem could be low water pressure coming to your home. Open a faucet elsewhere in the house and you'll know instantly. The pipes going to the washer could be blocked or broken. In the next step you'll be removing the feed hose from the faucet anyway. It takes just a second — and a bucket — to test the water pressure here.

If the washer fails to fill with water completely, or fills very slowly, the place to start looking is for clogged or blocked lines or filters.

The first job is to spot the sources of water. Normally this is quite easy to do. The hot and cold faucets connect to the two inlets on the back of the washer through two hoses. Often both ends of the hoses have filters to protect the machine itself and your clothing. Shut off the water. Put a towel beneath the hoses before you unscrew them (some water will leak out). You can now inspect the filters and clean them if this is necessary.

If the filter at the top of the hose (attached to the faucet) is clean, chances are good that the bottom filter is also clean. However, it takes just a few moments to be sure. It's possible that someone has periodically cleaned that top filter but has ignored the one closest to the machine.

The inlets on the washing machine lead to the double inlet valve and the solenoids, which open or close according to signals sent to them by the timer mechanism. Most often there are filters on this valve assembly. Once again, check these and clean them if necessary.

As mentioned earlier, the valve mechanism consists of two solenoids. (See Fig. 9–2.) A solenoid is a plunger operated by electromagne-

FIG. 9-8 Checking a hose filter.

FIG. 9-9 Water level control.

tism. The current to the solenoid causes the central rod to move. This either opens or closes the valve, depending on what is desired. (In the case of a washer, the inlet valves are normally closed. By activating the solenoid, the valves are opened and water flows in, after which another pulse closes the valve again.

A build-up of corrosion or other particles can prevent the solenoids from doing their job. They might not open all the way, or might not close all the way. The signs of wear are often obvious. If the valve assembly is quite old, you can suspect it, but still eliminate all other possibilities before going through the bother of disassembling the washer to check the valve.

Leaks can result from a variety of causes. Once again, knowing where water comes in and exits the machine will help you to spot the problem. A visual examination of these areas will often reveal the source of the leak.

If you can't readily find the leak, a major clue is the appearance of the water. Water with soap and suds has been in the tub. The problem could be a leak in that tub or the gaskets that seal it, but it is more likely to be somewhere in the drain system. If the water is clear, the problem is most likely in the intake, but it might be a leak in the tub.

A small quantity of water obviously indicates a small leak. Look for signs of leaking around water connectors or hoses that are just starting to wear out and that might have tiny holes. (The hot water side is more prone to this than the cold water side.) If the quantity of water is quite large, the chances are very good that a hose has come loose or has broken.

COMPONENT SERVICE AND REPAIR

Access panels to get at the interior wiring of timers, on/off switches, and cycle selector switching circuits can usually be removed with a standard screwdriver, a Phillips head screwdriver, or a nut driver. Almost always, the method and manner for removing the panels is obvious even at a quick glance. Sometimes clips are used as well as screws to hold the panels on tight. Don't force anything. If a panel doesn't come off readily, examine everything carefully. The panel might require a tug; on the other hand, it might be held in place by a screw or clip that you've missed, in which case the tug can cause damage.

Consult the wiring diagrams in your instruction manual. Quite often the appliance will have a complete wiring diagram taped or glued to one

FIG. 9-10 Motor and start capacitor.

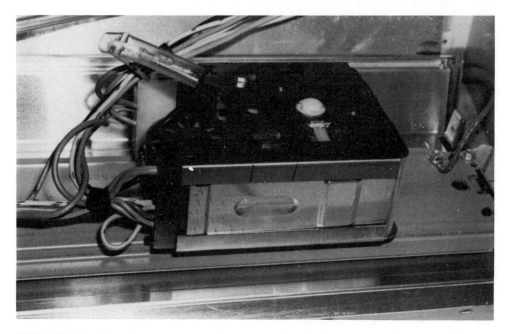

FIG. 9-11 Washer timer motor.

of the rear panels. With the power off, your VOM can be used to test for suspected open or shorted circuits or switches.

If a washer works well in all cycles except one, this usually indicates that either the selector switch or switch wiring circuit (including the timer) is bad, that the transmission or transmission linkage is not shifting into that particular cycle, or that one of the many motor windings has burned open or has shorted.

My own washer failed to spin and would leave the clothes soaking wet. Everything else worked fine. This could have indicated that the timer was bad and wasn't sending the signal to the motor and transmission. Or it could have meant that the transmission was unable to shift into high to spin the clothes. (Since the motor worked fine for the other cycles, that cause could be immediately eliminated.)

Hours could have been wasted tearing apart the washer to test the various parts. This is a perfect example of looking for the obvious and simple first. The entire fault was with the safety switch in the lid. If I hadn't already known that this was a possibility, I would have learned it by looking at the wiring diagram on the back panel of the washer.

Visual inspections and continuity tests with the resistance scale of the VOM will generally isolate the troublesome component. Also check timers and mechanical movements for bent or broken parts.

If the washer does not turn on at all, the very first thing to do is to be sure that you are operating the machine correctly. Often servicemen rush out on emergency calls to fix a nonfunctioning new machine, only to find out that the owner isn't flipping the right switch or turning the right knob (and hasn't bothered to read the owner's manual).

Next, look to see if the power supply cord is connected properly. Check the main fuse or the service entrance breaker box, and check to see if there are reset breakers on the washer itself (study the operator's manual).

Testing the various switches and other controls can usually be done while the component is still in place. Unplug the connectors going to the component (with the power off!) and test the suspected component for continuity with your VOM.

Replacing that component is usually easy. Most components are simply held in place by screws or bolts. (Be careful! The water and moisture around a washer increases the amount of corrosion. It's easy to snap off screws and bolts if you try to force them.)

Most washer control knobs are of the pull-off variety, fitting over knurled shafts. Be sure the controls are in the off position, grab the knob and pull it straight out. Replace by aligning the knob in its proper position and pushing it straight back on.

Further inside the machine the same disassembly rules apply. Consult the owner's manual for correct procedures. If a manual is not available, study the unit carefully to be sure you undertake only the disassembly necessary to gain access to the defective part. Take plenty of notes and make sketches.

Replace all damaged or worn out parts with exact replacement parts from the dealer for your particular appliance. If plug or cord replacement is necessary, be sure to use the same type and gauge of heavy-duty cord in order to safely carry the high amperage needed to run the washer's various cycles.

Mechanical components which require care and attention include the gaskets for the washer door and the door hinges. A misaligned door will impede correct cleaning and creates a safety hazard. A check of the owner's manual will usually reveal instructions for making minor adjustments in the door hinges after moving or after long periods of operation.

INTERIOR MAINTENANCE

Many owner's manuals suggest occasionally sanitizing the interior of your washer to help prevent the spread of illness among family members. This is particularly important whenever children are suffering from communicable diseases, or when there has been exposure to hepatitis, herpes, etc. This sanitizing procedure is also good for periodic freshening and deodorizing of the washer.

For the best effectiveness, sanitize the washer immediately prior to washing a load of clothes. Pour 1½ cups of liquid chlorine bleach in the bleach dispenser and close the lid. Use a warm rinse, normal agitate and spin cycle, and set the water level selector at small load. Run the washer through the shortest wash cycle on the selector knob. The washer will fill with water, dispense the chlorine bleach, agitate and spin the water out, leaving the washer clean and sanitized.

Even when you do not sanitize the machine, it will increase longevity of the appliance if you use a soft, dry cloth to gently wipe dry all interior surfaces at the end of each laundry day. Leave the lid up for at least an hour, to let the inside air dry and freshen.

WATER HARDNESS

Certain minerals in water create what is called "hard" water. The effectiveness of detergents is decreased by the presence of these minerals. And if soap is used, the minerals combine with the soap to form scum. This can make washed clothes come out looking yellow or grayish. Water is considered to be "hard" if it contains over three mineral grains per gallon.

Hard water washing can be improved by using exactly the correct amount of detergent for the fabric and load size involved, by using a water conditioner along with the detergent, or by installing a water softener.

The minerals iron and manganese lead to yellow or brown stains. Chlorine bleach can *increase* this staining effect. So if the water in your home has a heavy concentrate of either of these trace minerals, it is suggested that you use a perborate or oxygen bleach. If you do use chlorine bleach, you must add a water conditioner such as Calgon or Spring Rain. Precipitating water conditioners such as washing soda or Cloimalene form insoluble particles when they are added to hard water. It may be very difficult to rinse these particles out after the wash cycle.

DETERGENTS

Many environmentally conscious people now use non-phosphate detergents and biodegradable cleaners. However, use of non-phosphate deter-

Washer Troubleshooting Guide

Possible Causes	Solution
Problem: Nothing happens.	
Improper operation	Read owner's manual
No power	Check fuse or breaker, outlet, cable, and wiring
No water	Check faucets, hoses, filters, and valves
Bad selector switch	Test; replace
Safety switch	Close lid; if things still don't work test switch and replace if needed
Problem: Slow filling.	
Low water pressure	Test at faucet
Bad faucets	Test; replace
Bad, crimped, or clogged hoses	Repair or replace
Clogged lines or filters	Clean or replace
Bad valve assembly	Test; repair or replace
Problem: Poor draining.	
Bad or clogged drain hose	Repair, clean, or replace
Drain outlet too high for pump	Replumb
Weak or bad pump	Test; replace
Bad switch	Test; replace
Problem: Leaks.	
Loose or damaged hose	Repair or replace
Loose or damaged gasket	Repair or replace

gent may shorten washer life and lead to a few washing problems. Such cleaners can leave white streaks or a powdery residue on dark-colored items. Bright colors can fade quickly. The removal of stains is more difficult. Fabrics become harsh and stiff, and whites turn light gray or at least appear to be less clean and fresh. A non-phosphate detergent also has the tendency to erode any fire-retardant fabric finish. And these cleaners develop quicker and harsher stains both on the interior and exterior of the appliance.

To improve the quality of washing with non-phosphate detergents, pretreat stains and grease spots before loading clothes into the machine, use soft water, and use the hottest water permitted for the fabric. Be sure to use enough detergent, and dissolve non-phosphate detergents in water *before* adding to a load of clothes. Never pour the detergent directly on any clothing.

The so-called all-purpose detergents wash dirt out of clothing, and can be used for soaking, pre-washes, as well as the actual washing. Be sure to read instructions that come with the package before using any detergent.

In soft water, any detergent or soap may be used. Soft water generally means that a lesser amount of cleaner is required to wash clothes completely. In hard water, soap forms a scum. For this reason, use detergents if your water tests hard. Soap should only be used if you also add a water conditioner such as Calgon or Spring Rain to the wash water.

At the end of the cycle, be sure to thoroughly wipe out the interior of the tub and the enamel exterior surfaces with a clean, damp cloth. Then as a safety step, be sure that all switches are off. (You can also lengthen the life of the hoses and valves by shutting off the water at the faucets.)

DRYERS

HOW A DRYER WORKS

An electric dryer is one of the simplest appliances in your home. It consists of little more than a motor to spin the drum, a heater element to supply the heat for drying, and a timer to control the length of time of operation. The simplest units have a timer that signals both the motor and the heating element when to start and when to stop. Many dryers also have a cool down cycle, during which the motor keeps going but the heating element is shut off. This allows the clothes to cool and fluff.

In a dryer, hot air absorbs moisture, is vented away, and new air is drawn in, heated and gets vented away in its turn. Most often a small fan is used to force air movement. This fan is generally doing two jobs at

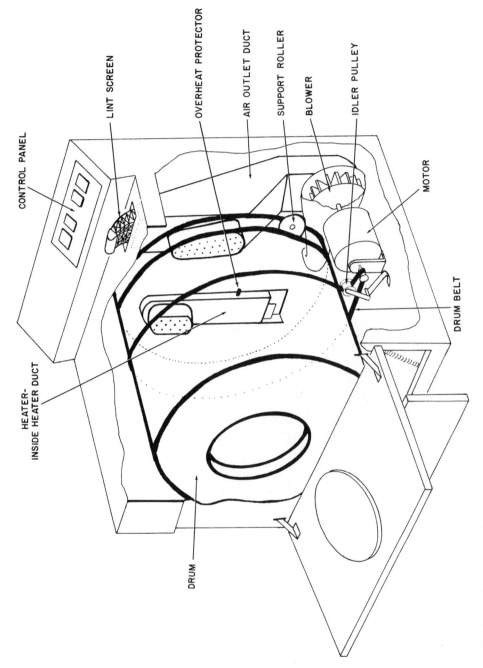

CONTROL PANEL

LINT SCREEN

OVERHEAT PROTECTOR

AIR OUTLET DUCT

SUPPORT ROLLER

BLOWER

IDLER PULLEY

MOTOR

DRUM BELT

HEATER-
INSIDE HEATER DUCT

DRUM

FIG. 9-12 Locations of parts of a dryer.

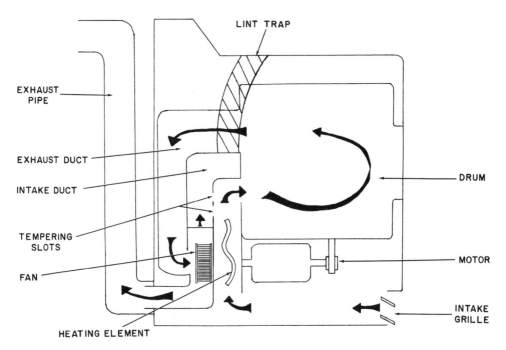

FIG. 9-13 Air flow in a dryer.

FIG. 9-14 Inside a dryer with the tub removed.

once. A part of it pulls in new air and pushes it through the ducting across the heating element and into the drum. Another part of it pulls air out of the drum, across the lint filter, and out through the exhaust duct.

As the clothes tumble in the drum, more surface area is exposed to the warmed and moving air. The tumbling action is further increased by blades inside the drum. A motor is attached to the drum, usually through some form of pulley and belt system. (The same motor is usually used to drive the blower fan for air circulation.)

The most common pulley and belt arrangement has a very long belt that goes completely around the drum, around the pulley attached to the motor, and also around an idler pulley. The idler pulley is spring loaded.

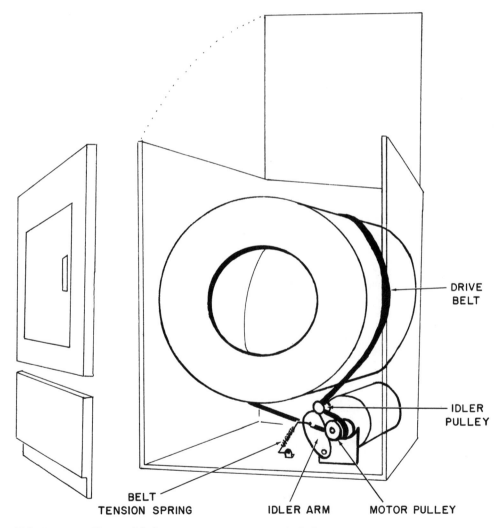

FIG. 9-15 Pulley and belt arrangement on a typical dryer.

Touch it with your finger and it will move. Its function is to keep the tension correct and to keep the belt in the proper alignment.

Some dryers have a belt and pulley at the rear of the drum. Smaller belts are used in this case. The idler pulley does double duty, keeping both under tension and in alignment.

The amount of heat and length of drying time is often determined by a timer mechanism. If the dryer has a cool down period, the timer will have two sets of contacts in it — one for the motor and one for the heating element. If both go on and off together, a single set of contacts is used.

More sophisticated units actually measure the wetness of the clothes in the drum, usually via contacts inside the drum. Since water conducts electricity, as the clothes touch the contacts a low current pulse keeps the dryer operating. The drier the clothes become, the less current they conduct; when dry, they do not conduct at all, and the dryer can shift into the cool down cycle.

THE THERMOSTAT

All dryers have a thermostat somewhere in the circuit. There are two basic types. One is an operating temperature thermostat that controls the

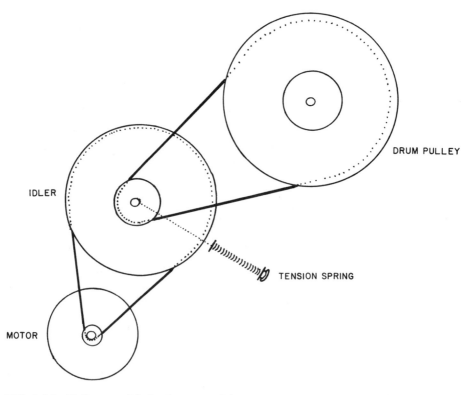

FIG. 9-16 Pulleys and belts in a rear-drive system.

FIG. 9-17 Dryer motor, pulley, and idler arm.

heater element during normal operation. Most dryers also have a high-limit thermostat, which prevents the heater element from overheating.

The operating thermostat is usually adjustable from the control panel of the dryer. It is virtually always set by a switch for preset ranges. By operation and construction, it is quite similar to the thermostat used in an electric range, with a bulb (usually located in the exhaust duct) and capillary tube.

The high-limit thermostat is most often found attached to the heater box. It is usually a fixed, bimetallic thermostat — a basic on-off affair. If the temperature rises too high, such as when the element is left on without the blower fan exhausting the hot air, this thermostat cuts the current flow to the element, protecting it, the dryer, and your home.

THE HEATING ELEMENT

The heater element is often wired through the centrifugal switch of the motor. This prevents the element from energizing if the motor isn't operating at the operational speed. It might also be controlled by relays. On the least expensive dryers it might be wired directly to the timer motor (although often through a relay).

FIG. 9-18 A high-limit thermostat.

Testing the heater element itself is done the same way as testing any heater element. Use your VOM to test for continuity. Remember to unplug the leads to it (with the power off) before testing. It isn't necessary to remove it for testing, although this is easy to do.

The element resides in a heater box, accessed through the back of

FIG. 9-19 The high-limit thermostat is usually on the heater box.

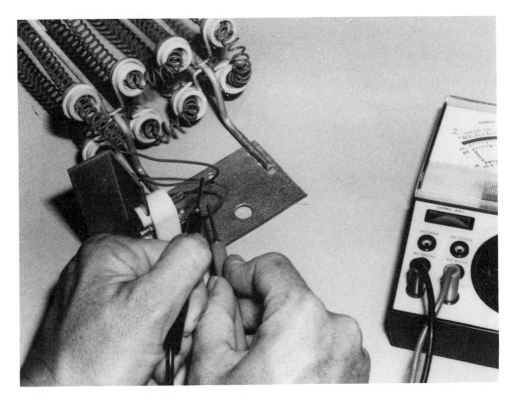

FIG. 9-20 Testing a dryer heater element.

the dryer. Getting it in and out can be a little tricky due to the legs and standoffs for the element. Take your time.

If replacement of the element is necessary, it's time to inspect the dryer. Heater elements do age and burn out. In many cases, there is a reason for the failure. Perhaps the lint trap isn't being cleaned often enough. Or the high-limit thermostat might not be working and is allowing the element to overheat on a regular basis. Since the back panel is off for the replacement, it takes only a few moments to inspect and check the thermostat. It takes even less time to clean the lint trap.

THE EXHAUST DUCT

Many people clean out the lint trap religiously but forget that the exhaust duct itself can become filled with lint. (Take a look outside where the vent comes out. After a relatively short time you'll find a layer of lint.) On a regular basis (every six months or so) clean out the exhaust duct. This will not only let the dryer operate more efficiently, it can also increase the life of the unit, and will even serve well in protecting your home from a fire hazard.

Dryer Troubleshooting Guide

Possible Causes	Solution
Problem: Nothing happens.	
Improper operation	Read owner's manual
No power	Check fuse or breaker, outlet, cable, and wiring
Bad selector switch	Test; replace
Safety switch	Close door; if things still don't work test switch and replace if needed
Problem: Dryer runs; clothes do not dry.	
Heating element bad	Test; replace
Fuse or breaker blown	Replace or reset
Thermostat bad	Test; replace
Bad timer	Test; replace
Bad switch or relay	Test; replace
Problem: Drum does not spin.	
Belt loose or broken	Repair or replace
Bad motor	Test; replace
Problem: Poor drying.	
Clothes excessively wet	Check washer
Too many clothes	Reduce load
Insufficient ventilation	Clean filter, ducts, etc.

WASHER AND DRYER BASICS

INSTALLATION TIPS

A home washing machine normally operates off a 110-volt ac wall outlet. Ideally, the appliance outlet should be on an individual 15-amp breaker or time-delay fuse at the home's service entrance. The dryer requires a 240-volt outlet and always must have its own circuit.

Washers come with a three-prong grounded plug. For safety, the cord should always be plugged into a three-hole grounded wall socket. If your home is equipped with the old-style two-wire outlets, any hardware or variety store can supply you with a conversion plug to adapt the washer cord to the two-wire outlet. The conversion plug has three holes to accept the washer plug, two prongs to slide into your two-hole outlet, and a short ground wire with a spade lug attached to the end. This spade lug must be slid under the metal screw holding the cover plate over the wall plug in order to make a solid ground connection. *Never* use a conversion plug without properly attaching the spade lug grounding wire.

For added safety, many of today's washers are sold with a separate external grounding kit. (The dryer plug and chassis already has this built into it.) This is a heavy wire that is connected from the metal chassis or frame of the washer to a solid electrical ground spot in the home (such as a metal clamp around a cold water pipe in the home plumbing system).

It is essential that the new washer or dryer be perfectly level so that it rests solidly on the floor. Set the appliance down on a bare floor, not on carpet. They work best on a concrete floor. Use a level, and adjust the appliance's leveling guides (or screws) on all four corners. Generally, you raise the corner height by turning the leveling screws to the right, and lower each individual corner by turning the leveling guides to the left.

After your level shows the machine is solidly in place (level side to side as well as front to back), then check for vibration as the washer spins. Turn the cycle selector to the spin dry mark, plug the cord in, turn the machine on, and make final adjustments in the leveling guides as necessary to keep the washer from vibrating. Occasionally, it may be necessary to place small wooden or metal shims under one or more of the corners in order to give the unit a stable footing.

After making the proper connections to the electrical outlet and to the hot and cold water intakes and the outlet drain, check the thermostat setting on your hot water heater. For best efficiency, the hot water entering the washer should be approximately 140 F. Adjust your hot water heater accordingly.

MOVING TIPS

Carefully take out any removable interior parts and pack them in a separate box. Moving interior components that cannot be easily removed should be securely fastened into place, using heavy tape or wire.

If there were shipping bolts and spacers provided with the original packing box, reinstall them. If the original packing material has been lost or discarded, check with the nearest local manufacturer's outlet to see if the service department can provide you with the necessary packing materials and information.

If you are storing your washer in an area of extreme cold weather, you must remove all water from the machine in order to protect it.

Disconnect all hoses from the input water supply. Put the machine on spin dry, and turn the unit on. Let the machine spin for a couple of minutes to let all of the water drain out. Turn off the machine, close the lid, and pull the electrical plug. Take the drain hose off of the pump outlet and drain water from both the pump and hose. Reconnect the hose to the pump outlet.

A washer or dryer should be stored only in the upright position. When moving them, even a short distance, keep them as upright as possible, and with as little jostling as possible.

EXTERNAL MAINTENANCE

For safety, always disconnect the electrical cord or turn off the service entrance breaker or fuse before servicing or even cleaning the appliance. There are many special manufacturer's cleaners and enamel paints, which may be recommended in the owner's manual for your particular appliance. Or check with any hardware or variety store for general all-purpose enamel cleaners. Never use gritty, harsh, or abrasive cleaners on any surface of a modern washer or dryer. Do not use oxalic acid or any commercial rust remover in the tub or near the top of the washer.

At the end of each laundry day, clean top and both sides of the lid by wiping with a damp cloth. Turn off both the hot and cold water faucets in order to reduce the pressure on the water hoses when not in use. The lint filter should be cleaned at the end of every washing or drying cycle. Consult the owner's manual for the proper method for cleaning the lint filter on your machine.

As needed, thoroughly clean all enamel, porcelain, and painted finishes with a soft, damp cloth. Commercial porcelain and enamcl renew spray finishes will work well. If the finish of the cabinet is accidentally scratched or marred, use a can of spray enamel for a color-matched touch-up.

Chapter 10
Ranges

Despite the dependability and wonderful versatility of today's electric ranges, this appliance is relatively simple to understand, and often easy to service. Heating elements usually plug in, which means replacing a bad element takes just a few seconds. With most ranges, virtually all parts that might need to be replaced are accessible without having to slide the oven out from the wall. Servicing the range requires little more than lifting a hood or panel, unplugging the old unit and plugging in the new.

HOW IT WORKS

The basic oven/range is a heating device, designed to convert electrical energy into heat energy. There are stove-top elements, and one or more heating elements inside an insulated cabinet (the oven) for baking, toasting, etc.

Temperature control is handled in several ways. The simplest is the use of a rheostat. The more the knob is twisted, the more current will flow through the element and the more heat it will generate.

More popular now is a switch with multiple discrete positions on it, five positions (warm, low, medium, medium high and high) being the most common. Most often this arrangement will use a double heater element. The switch allows either 120- or 240-volt current to flow to one or both of the elements. For example, on high, 240 volts is flowing through both elements; on medium high the 240 volts will flow only to the outer element and the inner element will be cut off. On medium both elements are provided with power, but only 120 volts each. For low, the inner element is cut off, and 120 volts flows only through the outer element. On

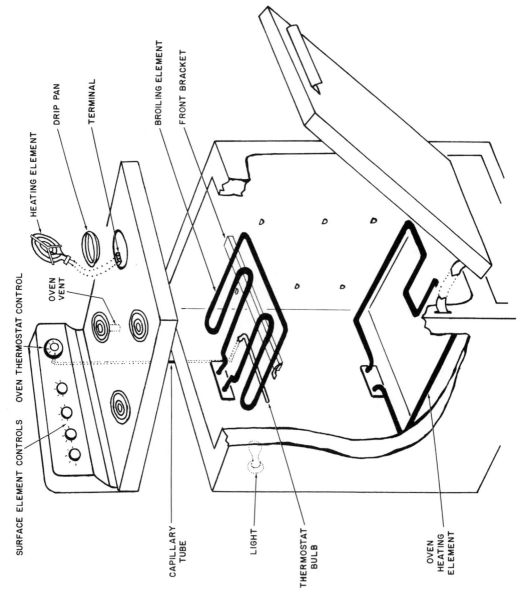

SURFACE ELEMENT CONTROLS   OVEN THERMOSTAT CONTROL

HEATING ELEMENT

DRIP PAN

TERMINAL

OVEN VENT

BROILING ELEMENT

FRONT BRACKET

CAPILLARY TUBE

LIGHT

THERMOSTAT BULB

OVEN HEATING ELEMENT

FIG. 10-1 Location of parts on a typical electric range.

warm both elements are fed from the same 120 volt leg, thus dividing the total voltage between them and supplying each with about 60 volts. Other arrangements can be used, depending on the number of settings desired. For example, you can easily provide an additional setting by sending 120 volts to one element and 240 to the other.

Another and newer type of temperature control has a bimetallic switch incorporated into the circuit. It is called an infinite heat switch. Depending on the temperature at which you set the knob, a bimetallic switch breaks the circuit more or less often. If the control is set for high, power is flowing constantly. In the medium position, it is flowing at full force half the time, and is completely shut off the other half. On low, power is flowing for only brief periods of time.

The owner's manual that came with the oven should tell you which system your particular range uses. Knowing exactly which will help in troubleshooting.

In the oven section, one or two heater elements are thermostatically controlled. A dial is used to set the temperature to a certain point, and the bimetallic thermostatic switch turns current on and off in order to maintain the inner oven heat at this level. This thermostat can often be adjusted. Normally this isn't necessary, however, and most manufacturers recommend that you leave the thermostat alone if the cooking temperature is within 25 degrees accuracy.

There are two basic schemes used for adjustable thermostats. One

FIG. 10-2   Inside a range control panel.

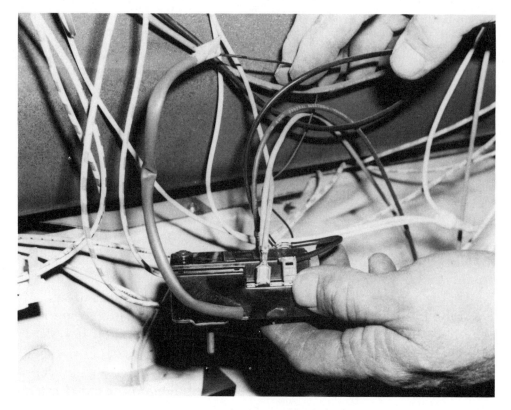

FIG. 10-3 An oven's thermostat and capillary tube.

has a movable arm on the thermostat. A set screw holds the arm in place. Loosening the set screw allows you to move the arm so that the resulting temperature is either higher or lower. Depending on your make and model, the little marks on the thermostat might represent 10 F each or 25 F each. Unless you have access to the literature specific to that thermostat, you probably won't know what the marks mean exactly unless you use an accurate oven thermometer.

A second method of adjusting the thermostat requires even more guesswork. With this kind a set screw on the shaft leading into the thermostat is loosened, and the shaft is rotated slightly in the appropriate direction.

Almost all oven thermostats are connected to a capillary bulb. This is a temperature sensing device that isolates the actual thermostat from the full heat of the oven. A bulb and small tube lead to the thermostat. For accurate readings, the tube must be between about ¾ to 1 inch from the oven walls. Closer or farther away will lead to inaccurate readings and could even result in an oven that is set at 400 F to heat at 500 F at one moment and 300 F at the next.

FIG. 10-4 Adjusting a thermostat.

The bulb and tube are filled with either a gas or a liquid. Helium is the most common gas used. Potassium hydroxide or sodium hydroxide are the most common liquids. Both of these liquids are dangerous. If you get them on your skin, wash immediately and thoroughly.

In a few top-quality ranges, the heating element consists of several coils, each with a different resistance to the flow of current. The more resistance, the hotter the element (coil) becomes. On many electric stoves, the surface burners glow red-hot when set at the maximum heat positions.

Occasionally, the heating elements will also be controlled by a timing mechanism or circuit. (The simpler models have a clock and timer, but these do nothing more than keep time and buzz.) If a true timer is incorporated with your particular oven, it probably handles the job of telling the rest of the appliance when to turn on and when to turn off.

DIAGNOSIS

Unnecessary service calls are expensive and quite frustrating. Even the home do-it-yourselfer can waste a lot of time by jumping in and disassembling an appliance before testing first to find out if service is really necessary. The problem is often something simple and obvious, or it may even be a matter of improper operation.

Since most heating elements in a stove work independently of each other, isolating a defective part is usually pretty easy. A single burner, timer, switch, or thermostat has failed or a mechanical part is broken, affecting the operation of the range.

Testing of individual parts is normally a matter of visual inspection, or through use of the VOM. Self-cleaning oven units generally involve the addition of a sensor, self-cleaning transformer, dual-range thermostat, and a hot wire relay. All may be VOM tested if not working properly. Be sure to use a note pad to indicate the placement of all connecting wires before removing a defective part, and use your notes carefully during reassembly with the new part.

Replace all damaged or worn out parts with exact replacement parts from the dealer. The model number of your appliance plus a description of the part should be enough to get an exact replacement.

If the range does not turn on at all, the first place to check is the oven itself for reset breakers. The owner's manual will show you where these are and what they do. (Most of the time, when there are breakers or fuses in the range itself, these are for various auxiliary outlets and devices.)

If there are none, or if the range breakers are okay, check the main fuse or breaker at the service entrance box. There will be two—one for each 120-volt leg of the total 240 volts. Some heating, but insufficient in amount, could be an indication that one of these two has blown.

Look to see if the power supply cord is connected properly. If possi-

FIG. 10-5 The range will have two fuses or breakers. Check them both.

ble, check the outlet box, the plug, and the wiring junction block inside. This means that you'll probably have to slide the oven out from the wall. Do so carefully, and only after you've shut down all power at the main box. There is a lot of amperage flowing when the oven is working. If the connections weren't made correctly in the first place, the contacts could become charred.

When one part of the range works fine, but another part does not— for example, if the surface burners all work just fine at all settings, but the oven doesn't do anything—the range is probably getting sufficient power. Double check the main fuses or breakers, just in case. Then once again shut off all power going to the appliance.

The surface heating elements usually plug into a socket. Examine the plugs, the sockets, and the wires going to the sockets. The whole problem could be that a removable element simply hasn't been plugged back in correctly. Burned or charred contacts is a sure sign of trouble.

Test the heating elements just as you would any heating element.

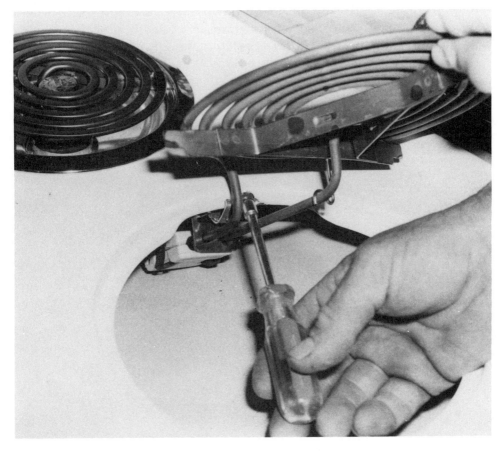

FIG. 10-6 Unplugging a surface heating element.

Disconnect at least one side and use your VOM to test for continuity across the element, preferably in the x1 range. A low reading is normal. (A reading of zero probably means that you're not doing it right.) If the reading is very high or infinite, it's time to replace that element.

A good way to get comparative readings is to test other heating elements of the same size. You can also swap elements of the same size as a part of the process of elimination. For example, if the range has two identical surface elements, with one working fine and another not working at all, or just barely, swap the two. If after the swap the suspected element works fine but the other known-to-be-good element is acting up, the problem isn't in the element but elsewhere in the circuit. In this case you can begin to back-track — test the switch, the wires, etc. until you find the true source of the problem.

If the oven fails to operate, determine if the oven selector knob is in the proper position to bake, toast, broil, etc. See if the temperature control is set correctly. If the range is a self-cleaning model, see that the "clean selector" is in the proper position for cooking instead of cleaning. On some automatic timer ovens, manual operation is not possible unless a "stop time" knob is in the right position. Once again, read the owner's manual, and then eliminate all the obvious possibilities first.

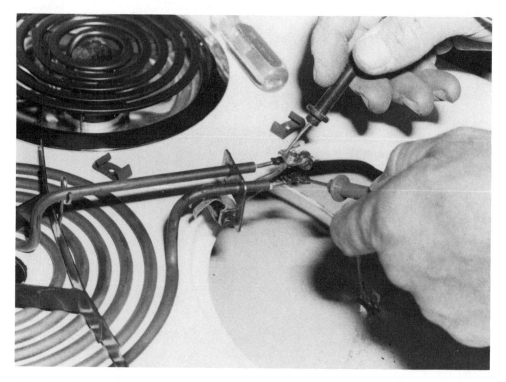

FIG. 10-7 Testing the element with a VOM.

Once you've eliminated operator fault, test the elements in the oven the same way as detailed above for the surface elements. Test for continuity. If possible, swap the suspected element for a good one. Then go after the switches, wiring, thermostat, and so forth.

Poor cooking results can be caused by a wide variety of reasons, almost none of which have anything to do with the stove itself.

Test to see that the range was installed straight and level. This can be done the traditional way, by using a builder's level of at least 12 inches in size. An even easier way is to simply put a small amount of water in a frying pan and set it on the various burners. If the oven is not level, the water will drift in the pan.

To make the oven level, turn the threaded legs in or out. Some will require that you first loosen a locking nut. If the oven is badly out of level, you may have to pull it out from the wall to get at the back feet.

Are the pans being used made of the materials recommended for the best results? Were all pans in the oven at least a couple of inches from the oven side walls and each other in order to allow for correct heat circulation during the cooking? Are the pans the right size for the recipe being used? Is the recipe reliable and tested? Are the ingredients fresh and the right kind(s)? Was the oven pre-heated if called for? Do your utensils have smooth, flat bottoms which make even contact with the heating elements?

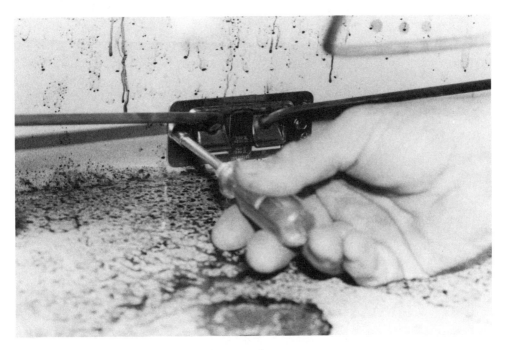

FIG. 10-8 Unplugging the oven's heating element.

Temperature control problems could also be caused by the control circuits and devices. Testing those devices might be in order once you've eliminated all the possible common and ordinary (and more likely) causes.

A visual examination is always first. The capillary tube leading to the thermostat might be broken. Wires or contacts might be loose, broken, or otherwise damaged. A slight discoloration is normal in most instances, but it is also a signal of a spot to watch in the future. Blackened contacts represent a real problem. The contact block and possibly the wires going to it will have to be replaced. Something has come loose and has allowed arcing to take place. Such problems should be cared for immediately, before ever applying power to the stove again.

Switches and thermostats both can be checked for continuity. (Thermostats are covered in depth in Chapter 6.) The thermostat can also be tested by placing an accurate oven thermometer in the oven. Let the oven thoroughly warm (at least 20 minutes) before checking the thermometer. (And don't forget to use heat insulated gloves or pads!)

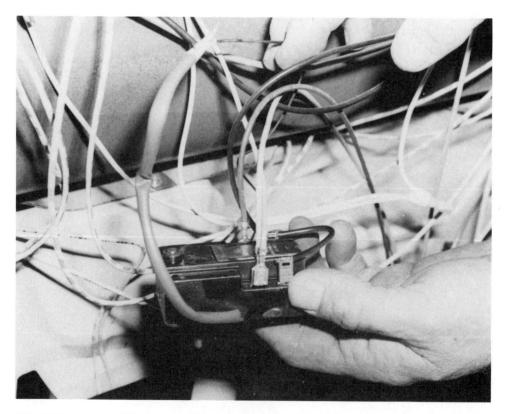

FIG. 10-9 Be careful when handling the capillary tube. The liquid inside can be dangerous.

Small differences can often be adjusted at the thermostat (see above). Larger differences indicate that either the element or the thermostat is about ready to be replaced.

DISASSEMBLY

Many of the problems with a stove can be completely diagnosed and fixed without ever having to move the oven away from the wall. However, there will come times when this is necessary.

Consult the owner's manual for correct disassembly procedures. If a manual is not available, study the unit carefully to be sure you undertake only the disassembly necessary to gain access to the defective part.

If you don't have a manual for your range, obtain one from the manufacturer. Good manuals contain wiring diagrams and instructions for the disassembly of sections needing more frequent servicing, repair, or cleaning.

There is heavy insulation in the door and walls for a specific reason. That insulation may even be installed in a very specific way. If your disassembly requires the removal of insulation, pay close attention to how it is to go back in place. Merely shoving it back in any old way is almost certain to cause problems later on — some of which can be serious.

Most stove control knobs are of the pull off variety, fitting over knurled shafts. A few have set screws that go through the knob and lock onto a flat portion of the control shaft. Carefully inspect the knob first. If the knob is of the set screw variety, loosen the screw with an appropriate screwdriver (small blade) and the knob will come off easily. If the knob fits over a knurled shaft or has a springy clamp, extra care is required. Be sure the controls are in the off position, grab the knob and pull it straight out. Don't force it. Replace by aligning the knob in its proper position and pushing it straight back on.

Be sure the stove is cool before removing any of the surface heating elements or trim rings, drip bowls, etc. Lift up the edge of the heating coil (½ to 1 inch) directly across from where the coil plugs into the stove. Unplug the element and lift it out.

The baking and broiling heater elements inside the oven are most often of the plug-in variety. Replacement of either is usually fairly simple. However, at times a further disassembly will be required. This is especially true when not just the element but its mounting contact is damaged.

Access in these cases requires moving the oven away from the wall. One or more back panels will have to be removed, and the insulation carefully moved aside.

With rare exception, you can quickly and easily disassemble everything involving an oven with nothing more than a screwdriver or nutdriver (and a bit of care).

The major task you face is that of moving the oven out from the wall so you can gain access to the back. Once you've done this, the rest is generally simple enough.

THE OVEN DOOR

There are a number of reasons that you may wish to remove the door to your oven. Perhaps it would make manual cleaning of the interior easier. Or you may have to take the door off in order to replace a broken oven glass window. Consult the owner's manual to see how to remove the door. A physical examination will also tell you just about everything you need to know for the job.

In general, there are slip-off doors, clip-on doors and screw-on doors. Some stoves use a combination of these, depending on where the door goes and what it closes.

To remove most slip-off doors, open the door to its first stop position (the broil stop — about 3 inches). Grasp the sides of the door near the top. Lift up and put on the door, pulling it away from the stove at the door bottom.

To replace this kind of door, fit the hinges into the slots in the frame and push the bottom of the door. The hinges should go into the slots evenly so there is no binding. Be sure the corners of the door set in as far as they will go after the door is back on.

To remove clip-on doors, open the door all the way. On many models, such as Whirlpool and some GE ranges, there are two clips on the door in front of the hinges. Swing the two clips so that they cross over the hinge slots. Other models have slide-out clips. (Consult the owner's manual and examine the door.) After the clips are slid out, close the oven door slowly until it comes to a stop. Grasp the sides of the door and lift it slightly. Tip the top of door toward the oven so the hinges move out freely, then pull straight out.

A more complete door disassembly will require removal of the back panel. Normally this panel is simply screwed into place. After the panel is removed, the door will not be as solid or stable, so move carefully.

Insulation in the door is put in place carefully during manufacture. Be sure to put it back in exactly the same way.

Simply loosening, but not removing, the screws to the inner panel will often allow you to remove the outer panel of the door. This is sometimes the easiest and safest way to gain access to the components inside the door, such as the tension springs.

Occasionally the door will not meet the main frame of the oven squarely. Before you go to the bother of aligning the door, first check for other possibilities. The gasket might be loose or damaged, or the door itself might be physically warped. To align the door, loosen the screws that hold the hinges in place. These are usually on the inner sides of the door. You can now move the door into the correct position. Then carefully tighten the screws again.

What appears to be an alignment problem may be the fault of the spring tension. Quite often access holes are provided to make the adjustment quick and easy. Other ranges will require that you remove panels, either on the door or on the main body of the range. A visual examination will let you know what has to be done. Once you've found the adjusting control, trial and error will show you which direction of turn is needed for tightening or loosening.

MAINTENANCE

The modern range/oven is pretty much self maintaining. Even if it is not of the self-cleaning variety, there's not much to be done to keep the unit operating perfectly.

The drip bowl under each surface burner is to catch spills and to reflect heat back up into pots for faster and more economical cooking. Washing the drip bowls frequently will keep them more efficient, and will also keep the range looking bright and new. These drip bowls can be removed and cleaned right along with the pots and pans used for cooking.

Note that if a drip bowl begins to turn blue or gold in color, the pots you are using may not be flat enough to be making solid contact with the heating element, or are too large for the burner size. The heat is going more down into the drip bowl, instead of being reflected back up into the pan.

Most heating elements are self cleaning. They will basically clean themselves during normal operation. All but large spills will burn away.

To keep the smoke in your kitchen at a minimum, wait until the burner is cool and wipe away the spill, if it is large. Do not immerse heating elements in water. In case of stubborn stains, use a scouring pad and scrub. Then allow the heat from your next cooking to burn the stain out.

Inside or out, don't let the stove get too dirty before cleaning it. This is especially important inside. A regular and thorough cleaning, including activation of the self-cleaning cycle, is important. Once grime is allowed to build up, the heavy scrubbing required to clean it up may actually damage the range surface.

Exterior Cleaning and Maintenance

Part	Cleaner to Use	Cleaning Method
Outside surface	Soft cloth; warm, soapy water. Nylon or plastic scouring pads for stubborn spots.	Regularly wipe cool range surface. Quickly wipe up spilled foods containing acids (citrus, vinegar, milk, etc.). They will mar finish. Never use abrasive or harsh cleansers or metal wire scouring pads.
Top of range heat elements	None required.	Spills, stains will burn off. Do not wash in water. Never place in oven during self-cleaning cycle.
Glass/ ceramic	Damp paper towel with cleaner/conditioner. Mild cleanser.	Cool before cleaning. Squeeze dab of cleaner/conditioner directly onto stains. Wipe with damp paper towel; remove excess cleanser with a fresh paper towel.
Control knobs/ chrome rims	Warm, soapy water; bristle brush.	Wash, rinse, and thoroughly dry. Do not soak in water.
Metal trim	Automatic dishwasher, or soapy steel wool pads.	Wash with cooking utensils. Do not place in self-cleaning oven.
Enamel/ porcelain drip bowls	Dishwasher or warm, soapy water.	Wipe off spills. Wash with other cooking utensils. Or place upside down on racks during self-cleaning oven cycle.
Teflon griddle	Do not immerse. Commercial stain remover.	Wipe after use. Frequently scrub Teflon finish with plastic scouring pad. Clean with mixture of 2 Tbs baking soda, ½ cup bleach with one cup water. Wash finish with solution. Wipe surface with salad oil before using again.

Part	Cleaner to Use	Cleaning Method
Broiler pan and oven racks	Dishwasher or warm, soapy water, or soapy steel wool pads.	Follow dishwasher instructions. Wash with other utensils. Do not place in self-cleaning oven.
Control panel	Warm, soapy water. Glass cleaner, if panel is glass.	Wash, rinse, dry thoroughly.
Oven	Commercial oven cleanser.	Spray and clean according to cleanser instructions. Wipe thoroughly.
Self-cleaning oven	Soapy steel wool pads or warm, soapy water.	Scrub stubborn spots. Rinse well with fresh water. Put strip of aluminum foil on bottom of oven to catch spills.

INSTALLATION TIPS

Several important considerations are involved in determining the location for a new kitchen range. Factors involved are convenience, safety, ease of maintenance and access, simplicity of cleaning, available exhaust venting, and the ultimate effect of the appliance on the home's heating/cooling system.

Most ranges operate on a 240-volt electrical system, as opposed to the 120 volts of the standard wall outlet. Breakers or fuses in the home's service entrance are usually rated at a higher current level than for the rest of the home—from 20 to 40 amps. For installation of a new unit in a new location, 240-volt electrical service must be made available, and the appliance must be properly grounded.

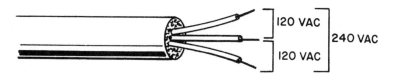

FIG. 10-10 Proper grounding is critical with a 240-volt appliance. Note that between the center wire and either of the other two is 120 volts, with 240 volts between the two outside wires. Improper grounding can mean that the body of the appliance will be "alive" with current.

Because most homes receive a 240-volt supply from the utility company, and this input voltage comes from a center-tapped transformer, the grounding is essential for safety and to meet building/electrical codes. With a center-tapped transformer, there is 120 volts from the grounded center tap to one end of the transformer coil, and another 120 volts from the center tap to the other end. (Usually the total service availability is from 100 to 200 amps, depending on your location and on the age of the home.)

The 240 volts available for range and air conditioning operation is provided by wiring from one end of the input utility transformer to the other. Improper grounding could result in a 120-volt "live" voltage potential between the metal parts of the range and the metal handle of a refrigerator or some other nearby 120-volt appliance. With bare wet feet, a user touching both appliances at the same time would be electrocuted.

Exhaust venting is necessary to remove both heat and odor. Smoke from the overcooking of greasy food is not only offensive, but can also stain walls and other interior surfaces. And during the hot summer, the exhaust vent helps keep kitchen temperatures at a tolerable level.

A stove should not be located in the direct path of air flow from an air conditioning or central heating register vent, since a flow of heated or cooled air moving across cooking utensils can drastically affect the time required to cook most foods.

If moving a range, at least a partial disassembly is recommended. The glass-paneled oven doors should be removed and packed separately, just as carefully as you would crate up fine china. Plug-in surface cleaning elements should also be removed. Use heavy masking tape to tape in place all knobs and switches. It is far safer to crate up the range in the same carton in which it was originally purchased. If that crate has been disposed of, the local dealer will usually be able to provide you with packaging materials and information at a very minimal charge.

SELF-CLEANING OVENS

The first time you have to clean an oven you start wondering if maybe humanity shouldn't go back to eating raw foods. Disguise it as the manufacturers may, even the best oven cleaners smell terrible and are unpleasant to use.

As an answer to this problem, oven manufacturers have come out with two different ways to make an oven self-cleaning. The basic idea is simple—the food is burned away so that nothing is left. This can be accomplished by either using a catalyst that causes the food to decompose at normal cooking temperatures, or by generating enough heat so that the food decomposes on its own. Hence the two methods — catalytic and pyrolytic (pyro means "fire").

A catalytic self-cleaning oven has a special surface on the oven walls. This surface is rough, to increase the total surface area, and it is also dark, to increase the effective temperature. This kind of oven basically cleans itself every time it is used.

Because it is so uncomplicated and cost effective, the catalytic self-cleaning systems are probably the most common today. The two major drawbacks are that the surface is rather delicate and that cleaning is often incomplete. Quite often, and especially with spills, a manual precleaning is required for the system to work at all.

Pyrolytic self-cleaning ovens have a special setting on the control knob. When you flip this knob to "clean," the temperature in the oven will climb to somewhere between 800 and 900 F, or about twice the normal cooking temperature.

To withstand these temperatures, the oven must have some special features. There is extra insulation in the oven walls and also various mechanisms to keep the high temperature under control.

For a pyrolytic system to work safely, the oven door must remain closed. If you open the door to an oven that is baking at 400 F, the wave of heat can hurt a bit, but probably won't do any real damage to you. Open that same oven door at 900 F and you're in for an instant and nasty burn.

To protect the owner, the doors on an oven that uses the pyrolytic system have an auxiliary motor that cranks the door tightly shut and keeps it shut until the temperature has dropped to a safe level. This is just one of the features that further complicates the oven.

Except for the various cleaning features and related parts (if any), troubleshooting of self-cleaning ovens is about the same as any other oven.

Troubleshooting Guide

Possible Causes	*Solution*
Problem: Nothing works.	
No power	Check fuse or breaker; check wires, cables, outlet
Problem: Light and clock work; only heats on low.	
Fuse or breaker blown	Check both breakers
Wire or connector bad	Check; repair immediately
Problem: Surface burners work, oven does not.	
Bad oven switch	Check; replace
Bad thermostat	Check; replace
Bad element	Check; replace
Problem: One surface burner doesn't work, or won't heat properly.	
Bad switch	Check; repair or replace
Bad element	Check; replace
Wires or connectors bad	Check; replace immediately
Problem: Poor cooking performance.	
Low incoming voltage	Test at outlet
Miscalibrated or bad thermostat	Calibrate; replace if needed
Wrong cooking utensils	Use the right ones
Vents blocked or dirty	Open or clean as needed
Problem: Self-cleaning oven won't clean.	
Improperly set	Read owner's manual
Insufficient heat or time	Set cycle correctly
Bad switch, wires, or connectors	Test; replace
Bad door interlock	Check; repair or replace
Oven walls too soiled or damaged	Clean more often; replace oven

Chapter 11
Dishwashers

The automatic dishwasher was invented in 1850 — 51 years before the invention of the first clothes washer, and 57 years before the first electric clothes washer made its appearance. Yet nearly 140 years later, a dishwasher is far less common than is a clothes washer.

A part of the reason that dishwashers have never quite gained the popularity of other appliances is that so many people have the attitude that an automatic dishwasher is an unnecessary convenience. There are some distinct advantages to having a dishwasher, however. Not only will the appliance clean your dishes more efficiently, it will at least partially sterilize them. The temperature of the water in the dishwasher is below the boiling point, but is sufficiently high (between 150 and 160F) and is in contact with the dishes long enough to kill the majority of the germs.

HOW IT WORKS

Proper loading of the dishwasher is critical to efficient operation. For example, if you load the bottom with a lot of pans and your dishwasher has only a lower spray arm, the dishes above may not be thoroughly washed. Worse, if a table knife slips down through the rack and blocks the motion of the spray arm, not only will the dishes be left unclean, you are likely to damage the dishwasher.

The soap dispenser holds a predetermined amount of dishwasher detergent. *Never* use regular dishwashing liquid in the unit. Dishwasher detergent is more alkaline than is regular soap, which helps it to remove food and grease. The detergent is also low sudsing to prevent clogging

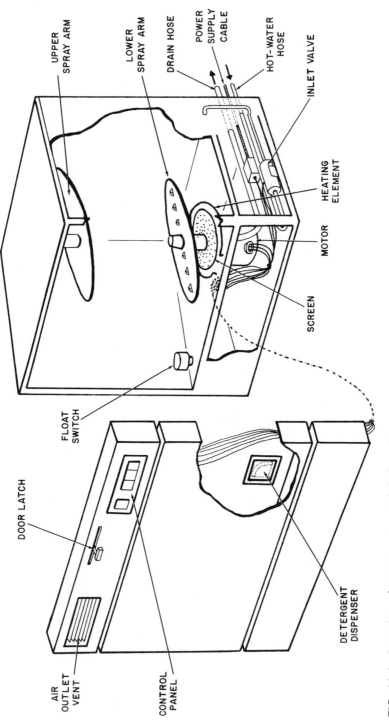

UPPER
SPRAY ARM

LOWER
SPRAY ARM

DRAIN HOSE

POWER
SUPPLY
CABLE

HOT-WATER
HOSE

INLET VALVE

HEATING
ELEMENT

MOTOR

SCREEN

FLOAT
SWITCH

DOOR LATCH

AIR
OUTLET
VENT

CONTROL
PANEL

DETERGENT
DISPENSER

FIG. 11-1 Location of parts in a typical dishwasher.

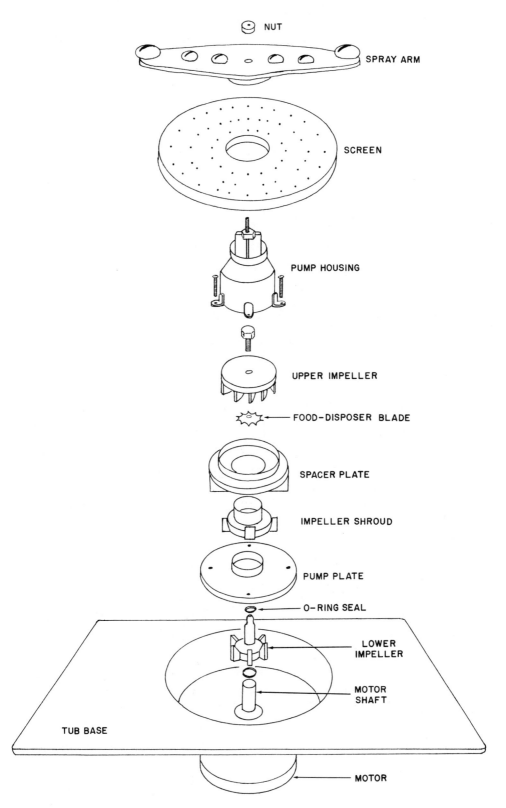

⊙ NUT

SPRAY ARM

SCREEN

PUMP HOUSING

UPPER IMPELLER

FOOD-DISPOSER BLADE

SPACER PLATE

IMPELLER SHROUD

PUMP PLATE

O-RING SEAL

LOWER IMPELLER

MOTOR SHAFT

TUB BASE

MOTOR

FIG. 11-2 The motor and pump assembly.

the dishwasher and to make rinsing easier. Some brands also contain a wetting agent to allow the water to flow off the dishes easier, thus preventing spotting.

Most dishwashers have a separate container for a rinsing agent for use with overly hard water. The receptacle for this second solution works much like the detergent dispenser. In many cases, a special concentrate is used, and the dispenser releases the proper amount in the final rinse cycle. The dispenser often holds enough of the concentrate to last for several months.

A timer knob or pushbuttons are used to control operation of the appliance. Simple units have a single knob, with as few as two settings — light wash and heavy wash, depending on how dirty the dishes are. Other dishwashers have pushbuttons that are used to select the washing cycle. This allows you to "program" the dishwasher.

In both cases, and particularly with the pushbutton models, the unit may have a rapid advance motor. This is an auxiliary motor attached to the timer to move it forward to whatever part of the complete cycle is called for, and to bypass those parts of the cycle that are not. In a few models the timer itself is a mechanical clockwork; in most it is a small synchronous motor.

Set into the door of the dishwasher is a safety switch, called an interlock. This prevents the dishwasher from working if the door is open, and thus keeps you and your kitchen dry (and safe). When you latch the door, the safety interlock is closed and the dishwasher can operate.

When the wash cycle is initiated, water flows into the unit. The knob or pushbutton opens the fill valve(s), and a certain amount of time is allowed for filling the tub. Solenoids are used to open or close valves to direct the flow of the water during the wash, draining, and rinse cycles. The solenoid consists of an electromagnet with a central moveable core. As the electromagnet is energized, the plug in the center moves, thus opening or closing the valve as needed.

The correct temperature for efficient washing is somewhere around 150 F, with 160 F usually considered to be the upper maximum and 140 F the minimum. (Some manufacturers list 150 F or 140 F as the maximum and 120 F as the minimum. It is known that the harsh dishwasher detergent at water temperatures of 150 F or higher can etch glassware.)

The water fills the bottom of the dishwasher tub. During the wash cycle it is kept hot by a heater element. Some units also have a built-in preheater to bring the water up to temperature, regardless of how hot the supply water is and to make up for loss in temperature in the pipes between the water heater and the dishwasher. (This temperature drop can be as much as 1 degree for every foot, which can lower the temperature below that which is needed for the dishwasher to operate at its best.)

A few units also incorporate a thermostat, which governs both the temperature of the water and when the unit begins to wash the dishes.

Often the same heater element that keeps the water hot also supplies heat to dry the dishes. The heater element in the bottom of the tub comes on. Heat, by nature, rises and goes across the dishes. The moist air is then vented, usually into the kitchen area. (A few units have a separate heating element for this function.)

Most dishwashers with this fast-drying feature have an "energy saver" switch on the front panel that allows you to disable the heating elements. The dishes then dry by normal evaporation.

A few models have a blower, usually mounted at the bottom of the unit. This blower forces air across the dishes for even faster drying. The blower is most often a separate and self-contained unit, complete with its own motor and an air duct leading from the blower housing to the dishwasher tub.

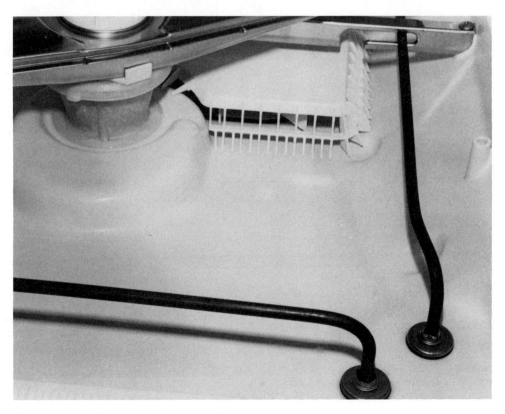

FIG. 11-3 A dishwasher's heating element.

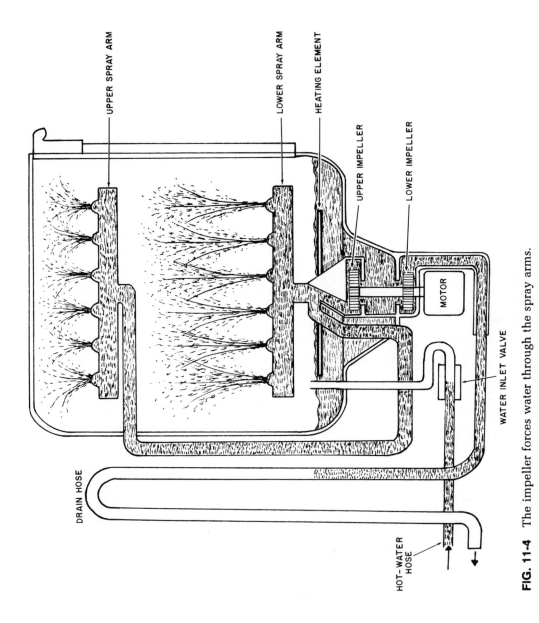

UPPER SPRAY ARM

LOWER SPRAY ARM

HEATING ELEMENT

UPPER IMPELLER

LOWER IMPELLER

MOTOR

WATER INLET VALVE

DRAIN HOSE

HOT-WATER HOSE

FIG. 11-4 The impeller forces water through the spray arms.

With the tub filled and at the right temperature, washing can begin. The main motor drives one or more impellers, and possibly a separate pump (although usually the impellers create the necessary pumping action).

Older dishwashers have an exposed impeller in the bottom of the tub. When the wash cycle is activated, the impeller spins, throwing water over the dishes with considerable force. Most newer models have an impeller hidden beneath a spray arm. The water thrown by the impeller is then directed at the dishes through small nozzles in the spray arm. At the same time, the spray arm spins for a more even distribution.

A few models have multiple impellers (usually two) with one supplying a lower spray arm and a second pushing water under high pressure to an upper arm.

All this spinning motion and driving power is generated by a $\frac{1}{4}$ to $\frac{1}{2}$ horsepower motor, usually of the split phase variety. (See Chapter 5.) This motor serves to drive the impeller(s) and any pumps used by the appliance. (As mentioned earlier, the timer is often a small synchronous motor. There may also be a third motor, for rapid advance of the timer mechanism.)

The main motor is usually reversible. This allows the same motor to drive the impellers in one direction to spray water over the dishes, and in the opposite direction to pump water out of the appliance. A reversible motor will have four leads. One of the leads is the common. One

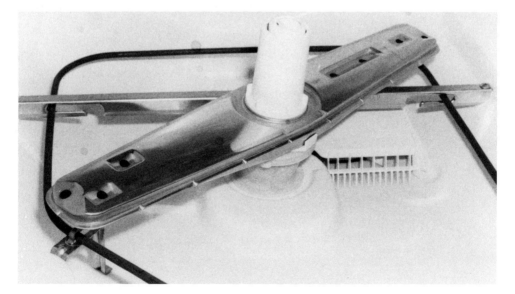

FIG. 11-5 The spray arm.

goes to the main running winding. The other two go to the two start-up windings that determine direction of spin.

Sometimes the motor and pump assembly will be attached in such a way that the entire motor/pump unit must be removed as a whole to work on either the motor or on the pump. Carefully inspect your dishwasher.

Pump or impellers are usually located at the bottom of the dishwasher and in the center. In some models the motor and pump are set off to the side. When located elsewhere, there are hoses leading to and from the pump assembly. (Whenever you have to get inside, visually inspect these hoses and all gaskets for leaks and corrosion.)

TROUBLESHOOTING

As always, the first step in troubleshooting is to look for the obvious. With dishwashers this becomes even more important since what seems to be a major malfunction is probably nothing more than something like old dishwasher detergent. A dishwasher that blows the fuse or trips the internal protective circuitry could have nothing wrong with it other than poorly loaded dishes that are preventing the spray arms from spinning.

Before anything else, get out the owner's manual and read it through again thoroughly. Many of these manuals give the owner detailed information specific to that brand and model, including troubleshooting tips.

Review the proper method (or methods) of loading the machine.

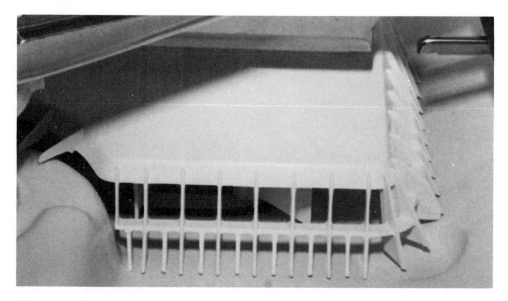

FIG. 11-6 The pump cover.

Probably the single greatest "malfunction" comes from users who shove in the dishes willy-nilly without ever looking in the owner's manual for the correct way to do it.

Next, eliminate as many obvious things as possible. If the dishwasher is completely dead, it should be an easy matter to check the fuse or circuit breaker. (You did mark the service entrance panel, didn't you?) If you have a portable dishwasher that plugs into an outlet, take a moment to see if power is getting to that outlet.

THE SAFETY SWITCH

On many units, a safety switch is mounted in the front panel and connected to the door latch. If the door isn't completely closed and properly latched, this switch is open and nothing is going to work. That is its function. But if the switch has gone bad, it won't matter if the door is latched or not. In this case, the dishwasher will run whether the door is closed or not, or it won't run at all. It's usually a simple matter to test this switch. An examination of the door and latching mechanism will generally reveal where it is and how to get to it.

With cabinet-installed units, it's important to keep in mind that there might be power flowing to the appliance even though it is not working. Even if you're certain that there is no power going to the safety switch, or to anything else you will be testing with your VOM, assume that there *is* power. Proceed cautiously.

If you have a portable unit, unplug it. If the unit is built-in and disconnection is difficult, remove power by taking out the fuse or flipping the breaker. The dishwasher is doubly dangerous in that it has both electricity and water going to it. Don't take any chances!

Examine and test any and all other switches in the same way. (For now don't worry about those automatic switches that are buried deeper inside the unit.)

THE TIMER

All dishwashers have some sort of timer mechanism, either a clockwork type or a small synchronous motor (or it might even be electronic). Regardless of type, repair of the timer is rarely possible. If the timer is malfunctioning, the solution is to replace it.

On most units the timer is mounted in the door or in the front panel. Access usually isn't too difficult. If any gaskets have to be removed, pay close attention to how the gaskets came out so that you can reinstall them correctly. Otherwise your simple test could turn into a major leak.

Examine the timer closely. Look for visual signs of corrosion, burn marks or other discoloration, bad contacts, and so forth. Depending on the model you own, you might be able to test the timer unit with a VOM. The easiest test of the timer is to test it in operation. It's rare for the entire unit to fail at the same time. More often, parts of the timing cycle will work while others do not.

Before ripping out the timer unit and replacing it, think of what happens during that part of the cycle. Also make note of any other features or functions that don't seem to be working. For example, if everything seems to work but the drying cycle, the fault might be in the timer, but it might also be in the heater element. If the dishes don't seem to be as clean as they used to, that could be another indication of a faulty heater element that is failing to keep the water hot enough.

With water involved, contacts tend to corrode faster, both on the timer and elsewhere. Before replacing any component, and anytime you get inside the machine, check all plugs and contacts, especially those internal and going to the motor.

THE MOTOR

Also look for internal circuit breakers or reset switches. Many motors have these overload protectors built in. That table knife that dropped down through the basket and caused the motor to seize could have tripped the protective circuitry. The motor itself might be (or might not be) just fine, with your only job being to push that reset button.

If the motor hums after reset, check the centrifugal switch mounted to the motor's casing. (See Chapter 5.) This is often a sign that the main winding is getting current but that the start-up winding is not. If the start-up winding doesn't energize, the motor won't start spinning—it just sits there and makes noise.

Loud humming of short duration, followed by a blown fuse or breaker is usually a sign of binding. A utensil might have dropped down to block the spray arm or impeller. The shaft itself might have been damaged by some careless earlier repair attempt. Or the problem could be inside the motor.

Look for the obvious first. If someone else was using the machine when the binding took place, try to find out if the cause was something simple before you spend the time and effort to tear apart the machine. If possible, cure the problem, reset the breaker, and try again.

As mentioned above, some motors have built-in resetable protective circuitry. If resetting the main breaker or replacing the main fuse doesn't

FIG. 11-7 Dishwasher motor.

work, you may have to disassemble the dishwasher to get at the motor. Quite often this can be done easily through a bottom panel.

THE PUMP AND IMPELLERS

Damaged impellers can cause uneven washing. Exposed impellers are particularly vulnerable, but any damage is almost always due to a careless operator. The easiest repair, as always, is making sure that the repair isn't needed in the first place.

The impeller might be screwed into place on the motor shaft, or it might be force fit. Both are prone to corrosion, which can make removal of the old impeller difficult. A corrosion cutting compound such as WD-40 can be used to help loosen a stuck screw or impeller. Move slowly and take great care not to damage the motor shaft. If you do, you'll probably have to replace the motor as well as the impeller or pump assembly.

Keep track of washers and spacers. These are there for a reason, and are usually in a specific order. Notes and sketches will help you reassemble everything in the proper order. Also pay close attention to any instructions that come with the replacement part(s).

Quite often, replacing a defective component will mean replacing all the gaskets at the same time. Removal of the pump assembly, for example, might easily result in damage to the gasket that seals the water inside the tub. Gaskets are inexpensive, especially when compared to the cost of replacing the motor and other parts when a poor gasket seal creates water damage.

Most modern units have spray arms instead of open impellers. This kind pumps the water through the arm, causing the water to spray on the dishes and the arm to spin. Some units have spray arms above and below; others have an arm that extends during operation.

As efficient as the spray arm is, it is also prone to clogging. Fortunately, with most units, removal and cleaning of the spray arm is fairly easy. It is attached to the motor shaft, which means that once again you have to be careful to not bend or otherwise damage that shaft.

LEAKS

There are two primary causes of leaks. One is a corroded water feed or drainage line. The second is a worn or damaged gasket. Both are easily checked, and quite often you'll find that the entire problem is nothing more than a build-up of sediments on a water sealing gasket.

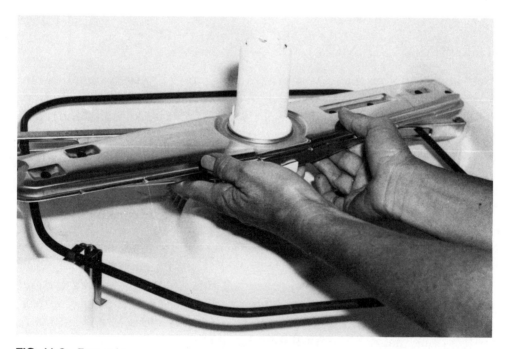

FIG. 11-8 Removing a spray arm.

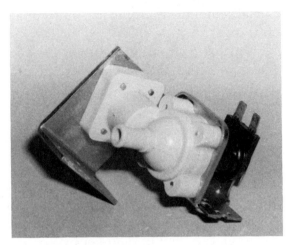

FIG. 11-9 Dishwasher water valve.

More often than not, a gasket leak occurs in the gasket that seals the door to the unit. Front-loading machines are considerably more prone to leak here than are top loaders.

Your first job is to visually inspect the dishwasher. Simply looking at the water from the leak can indicate where it is coming from. Quantity and appearance are both important. Clear water is a good indication that the water intake is at fault. Sudsy water means that the leak is more likely to be internal or with the drainage output. Small quantities indicate either pinhole leaks in the various pipes or a failing gasket. Large quantities indicate a broken pipe or hose.

The cause of the leak will usually be apparent after a visual examination of all hoses, clamps, filters, gaskets, seals, and so on. If you can't find the actual source right away, you should be able to track it backwards just by the appearance of the various water-carrying components.

At times the only real problem is a lack of cleanliness. A dirty gasket, especially the one that seals the door, may not seal properly. The inlet or drain lines might be clogged. These lines often have fine-screen filters in them. Theoretically, these filters are self-cleaning. In practice, they can clog almost completely from a combination of food particles and mineral deposits.

The filters are usually hidden, and getting at them almost always calls for a partial disassembly of the machine. As a general rule, there will be a filter in the pump and a filter inside any hose where it attaches. Clean running water will usually take care of the job, although a toothbrush can make the job easier.

Be sure that you've thoroughly sealed the connection. If need be, replace the gasket or washer that makes the seal. You may even wish to use a small amount of plumber's putty or another water sealing compound on threads.

Pay extra attention to any hose or connection that carries hot water. These are more prone to leaking. And, because hot water tends to carry more dissolved minerals than cold water, the hot water lines will tend to exhibit more corrosion.

POOR WASHING

If your dishes don't get clean, it's possible the problem is with the timer (especially if the dishwasher is old) or the water temperature or quantity. But the chances are better that the entire fault is in how you load the machine, the kind of detergent you use, or the hardness of your water.

Before tearing the machine apart, look for the simple and obvious. Read through the owner's manual. Learn how to properly load the machine, and what kinds of dishes, pans, and utensils should be loaded where. Examine the detergent you are using. Try dissolving a couple of teaspoons in a glass of hot water. It should dissolve quickly and completely, without leaving any residue. If the detergent is old, replace it. You may also want to try another brand of detergent. Still another problem involving the detergent concerns the hardness of the water. More

FIG. 11-10 Some dishwashers have a telescoping wash tower that rises during operation. Be sure that you don't block this or any moving parts, or interfere with the function of parts.

detergent is needed with harder water; less with softer water. The first case could leave you with uncleaned dishes, while the second could cause a film of detergent on the dishes.

It's possible that you'll need a rinsing agent for spot-free cleaning. (The owner's manual will tell you what type is needed.) This rinsing agent decreases the surface tension of the water, allowing it to run off more easily, and thus dry without spotting.

Many manufacturers recommend a vinegar bath to help solve film problems. To do this, put the dishes in the dishwasher, but without any detergent. After the machine has been running for about 15 or 20 minutes, shut off the machine, open the door, and pour 2 cups of white vinegar into the water in the bottom of the tub. Now complete the wash cycle.

This should remove the film. If it doesn't, you can use ¼ cup of citric acid crystals in the same manner. If the problem still isn't solved, the glassware is probably permanently etched due to the combined effect of the powerful detergent and excessively high water temperature.

You can prevent such problems by getting a water softener, by reducing the amount of detergent used, by reducing the temperature of the water at the water heater, or by a combination of these things.

Temperature of incoming water should be some where around 150 F but no higher than 160 F and no lower than 140 F. If the water temperature is below 140 F it won't dissolve grease well. (Touch lower edges of door on a front loader to feel for grease. If you find some, the water temperature is probably too low.) If you suspect that poor washing is due to low water temperature, draw a glass of hot water from the faucet nearest to the dishwasher and test it with a thermometer that can measure fairly accurately in the 150 F range.

You can increase the efficiency of the cleaning, and reduce service problems, by proper precleaning. Large pieces of food should be removed—these are guaranteed to cause trouble. Even if the dishwasher has a built-in garbage disposer or feeds the drain water through your household disposal unit, the more food that remains on the dishes, the greater the chance that cleaning will be insufficient and that clogging will be a problem.

It's best to quickly rinse off the dishes before placing them in the dishwasher. Better yet, scrub away all food that remains on the dishes, and use the dishwasher for the final cleaning and sterilizing. Not only will the dishes come out cleaner, you'll be adding years of troublefree service to your dishwasher.

With all these possible causes of poor washing tested and eliminated, you can turn your attention to the more complex (and more rare) causes.

Just as the water line filters can become clogged, so can the spray arms. A visual examination will reveal if this is the problem. The vinegar or citric acid bath mentioned above for filming problems might cure the problem. You can also manually clean the spray arms—but carefully! Never attempt to clean the orifices with anything that is harder than the material of the spray arm. A toothpick might work, but be very careful. If you enlarge the opening during the cleaning, you'll only be creating a new and possibly worse problem.

Dishwasher should fill at a rate of about 2 gallons per minute. Check the filters, and then the plumbing, if fill time is overly long. As mentioned in the previous section, the filters keep particles of hard material from getting into the pump. They can clog, thus reducing the washing action.

The easiest way to start testing a filling problem is to open a faucet near to the dishwasher. Observe the amount of pressure available. If it seems unusually low, the trouble could be temporary with the main water supply to your home. Try again later. Or you could have a major plumbing problem, such as old pipes. In this case the only solution is to replace the plumbing, which usually means calling in a professional.

The detergent dispenser is usually operated by a flip latch that is connected to a solenoid which in turn is connected to the timer. If you find a clump of detergent in the cup after a washing, your first suspect should be the detergent itself. Try a new box, or a different brand.

If the door itself has remained closed, something else may be wrong, but don't tear the dishwasher apart yet. Give the dispenser a thorough cleaning. Manually move the door to check for binding. If these things don't cure the problem, you'll probably have to disassemble the dispenser, which usually means disassembling the door. Visually inspect the latch mechanism. Work it manually a couple of times. If the mechanical action seems to be fine, the problem is either in the timer mechanism (test it) or in the solenoid (test it).

PREVENTIVE MAINTENANCE

There is very little that can be done in the way of maintenance on a dishwasher. Even the regular cleaning that is required on any appliance isn't needed very often on a dishwasher.

By the nature of the machine, it cleans itself inside each time you use it. Only rarely will there be a need to clean the inside, and then usually only if there is another problem. For example, if the water temperature has been too low, and the dishes you load in too dirty, the tub might develop a layer of grease.

You may find it necessary to use the vinegar or citric acid crystals for a more thorough cleaning if mineral deposits are a problem. But again, this indicates that there is a problem else where that requires your attention (the water is too hard, or the detergent is bad).

If interior cleaning is necessary, first try to find out what is causing the problem and cure it. After that, a few normal washing cycles should take care of the cleaning for you.

For exterior maintenance, wipe off the outside of the dishwasher lightly and often, rather than seldom and hard. If you keep ahead of the dirt, grease, and other deposits, they are much easier to clean off and you'll reduce the chance of damaging the surface by hard scrubbing.

While you're cleaning the outside, take a moment to wipe down the door gasket. Use only a clean cloth for this—not the one you used to wipe away deposits of grease from the outside. On a regular basis, visually examine all gaskets, seals and filters that are easily accessible. Clean or replace them as needed.

Troubleshooting Guide

Possible Causes	Solution

Problem: Machine fails to start.

Door partly open	Close securely
Bad switch/timer	Repair: replace
Open circuit	Check fuse or breaker

Problem: Dishes are not completely cleaned.

Bad soap	Use fresh detergent
Wrong soap	Use detergent recommended by the manufacturer; try another brand
Improper loading	Follow operation instructions
Dishes not properly pre-cleaned	Thoroughly rinse and pre-clean
Low water temperature	Adjust thermostat on water heater
Insufficient water	Check feed hoses for clogging and to see that valves are sufficiently open
Poor drainage; strainer clogged	Remove and rinse with tap water; preclean dishes
Timer bad	Test; replace
Solenoid coil bad	Test; replace

Problem: Water does not stay in supply tank for washing.

Drain valve leaks	Tighten valve flange; replace gasket
Inlet valve too low	Adjust, repair, or replace inlet solenoid

Problem: Machine noisy.

Improper loading	Load machine properly
Vibration; machine not level	Set machine solidly; adjust any level controls
Motor loose or out of alignment	Realign; tighten motor mounts
Core of one of the solenoids is not centered in coil	Realign core or replace
Impeller hitting impeller screen	Check and adjust

Troubleshooting Guide, *continued*

Possible Causes	Solution
Problem: Door or cover will not close properly.	
Binding/jamming	Loosen screws and readjust hinges and any seals or gaskets so door closes tightly
Bad gasket/seal	Replace
Problem: Insufficient water, or draining not complete.	
Low water pressure	Check water pressure from faucets; check machine inlet
Lines plugged	Clean hoses, filters, and drain valves
Solenoid bad	Repair/replace solenoid
Problem: Dishes don't dry.	
Incorrect water temperature	Set thermostat to 150 to 160 F
Leaking inlet valve	Replace washer
Faulty heating element	Set timer to heating cycle and see if element is heating; if not, check both timer and heating element; repair/replace
Problem: Tarnish on cleaned dishes.	
Improper loading	Load correctly
Improper detergent	Replace; try another brand
Too much detergent	Reduce amount of detergent
Insufficient rinse agent	Refill dispenser
Hard water	Use water softener or water softening detergents
Water temperature too low	Set thermostat at water heater
Metallic mismatch	Don't wash copper utensils with silver; don't let aluminum utensils touch other objects; don't let silver touch stainless steel
Problem: Glass etching.	
Water temperature too high	Lower temperature at water heater
Detergent too harsh	Use less detergent; try another brand

Chapter 12
Air Conditioners, Heat Pumps, and Electric Furnaces

Modern air conditioning systems have come a long way. Today, a lot of attention is being paid to moisture controls as well as temperature. Stirred by increasing energy costs, better and more efficient units are being built. Heat pumps, capable of providing both cooling and heating, are becoming more popular and are replacing separate units, often with a great savings in power.

There are very few climates where at least some heating and/or cooling isn't required. Even if you happen to be one of the few who doesn't use the heating/cooling system often, your home almost certainly has all of the components in one form or another.

HOW IT WORKS

An air conditioner or heat pump works by expanding and contracting gases, and by the movement of heat from one place to another. As a gas is compressed, it generates heat. As it expands, it creates cold and is capable of absorbing heat. This simple principle is what is behind all refrigeration units, whether it be the refrigerator or freezer in your home, or the air conditioner.

The tubes of the air conditioning unit are filled with a gas. This gas is usually either Freon-12 or Freon-22. The Freon gets squeezed and condensed into a liquid state by a compressor and a high-pressure condenser. It expands into a gas (evaporates) alternately in the low-pressure evaporator side of the system. To cause the high pressure, a compressor is used to literally squeeze the Freon gas. Heat is drawn away by a condenser fan blowing across the coils. The heat dispersion is further in-

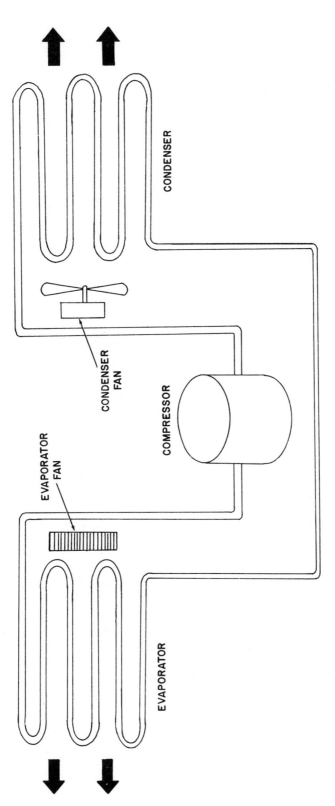

FIG. 12-1 A typical refrigeration system.

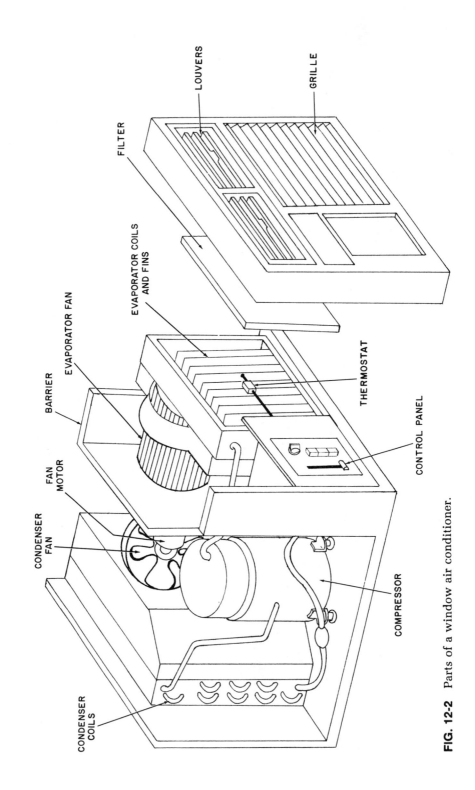

FIG. 12-2 Parts of a window air conditioner.

LOUVERS

GRILLE

FILTER

EVAPORATOR COILS
AND FINS

THERMOSTAT

CONTROL PANEL

COMPRESSOR

EVAPORATOR FAN

BARRIER

FAN
MOTOR

CONDENSER
FAN

CONDENSER
COILS

creased by the use of thousands of metal fins. It's important to keep these fins clean (but clean them gently), otherwise the heat cannot be effectively transferred away.

As the liquid passes through the tubing, it not only loses heat, it loses pressure. The low-pressure side, called the evaporator, turns the liquid into a gas, which cools as it expands. The tubes get cold, and another fan blows the cold air where it is wanted.

This is obviously a closed system. If the Freon leaks out, the system will become less and less efficient and will eventually stop. It's very possible that the compressor will be damaged or destroyed as a result. The compressor itself, which looks like a large cylinder in most units, is hermetically sealed. Repair of the compressor itself is not possible. If the compressor fails, replacement is the only valid solution (although the old compressor might have a trade-in value).

The air movement of the home is also a relatively enclosed system. This allows the air-conditioning system to also work as a dehumidifier. The inside air moves through the air return and across the evaporator (cooling) coils. This drop in temperature also condenses out much of the humidity. The cooler, dryer air is then blown back into the house.

In essence then, the system has two overall parts—the coolant sys-

FIG. 12-3 An air conditioner compressor.

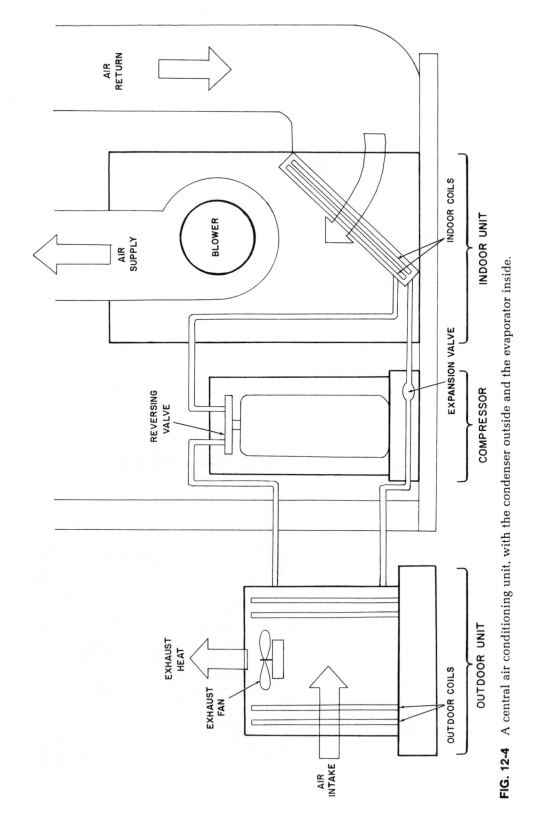

FIG. 12-4 A central air conditioning unit, with the condenser outside and the evaporator inside.

tem and the air handling system. The coolant system consists of a compressor, a high-pressure side (the condenser), a fan to blow away the heat, usually some kind of inline filter and drier, usually some kind of sight bubble, a capillary tube or loop, a low-pressure side (the evaporator) and the gas inside. The air handling side consists of a fan to move the air, the incoming and return air ducting, and at least one air filter.

On the simplest units, including all window air conditioners, all of this will be in a single chassis located outside the home. Other units will be split, with the condenser located outside and the evaporator located inside, often in the attic.

To control everything, at least one adjustable thermostat will be located in the home—usually close to the air return. This thermostat is almost always of the bimetallic variety. As the temperature inside reaches the set level, the contacts turn the compressor and fans on. Once the air is cool enough, contact is broken and the system is shut off.

The thermostat circuit is quite often supplied by a low voltage (24 volts ac) transformer, usually located within the air conditioning cabinet. Because of the low voltage, very small wires can be used in the walls, and any danger to the homeowner is greatly decreased.

Most whole-house systems have a setting to keep the house fan on all the time. This helps to keep the air in the house moving. Warm air

FIG. 12-5 The thermostat assembly.

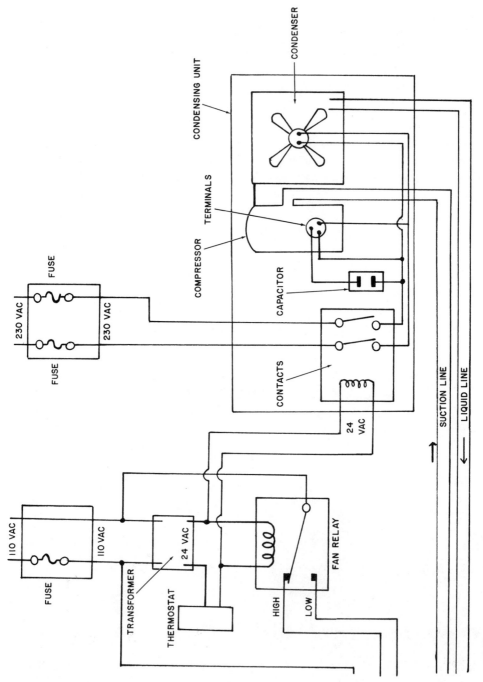

FIG. 12-6 Wiring schematic of an air conditioner.

rises, while cooler air drops. The air movement from the fan keeps the air in the home mixed. Especially during the hot and humid season, leaving the fan on constantly can actually reduce costs while making your home more comfortable. It also tends to put less wear and tear on the overall system since it doesn't have to cycle as often.

HEAT PUMPS

A special type of cooling and heating system is called the heat pump. This central unit takes heat from outdoor air (or water from a radiant system) and moves this warmth indoors during the winter. The cycle is reversed during the summer, with the unit absorbing indoor heat and disposing of it outdoors (or back into the water).

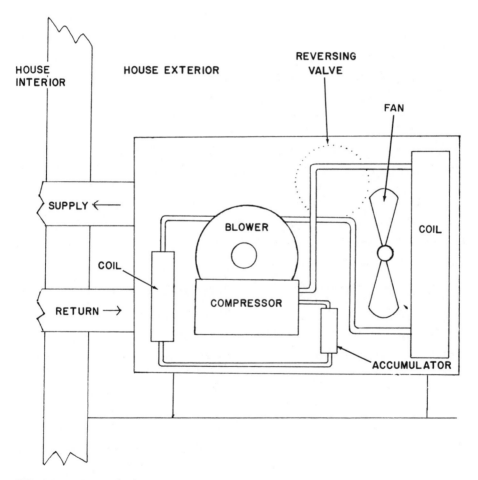

FIG. 12-7 Parts of a heat pump.

In general, a heat pump consists of an outdoor coil, indoor coil, and a compressor. When hot, the indoor coil picks up heat from the household air and ejects it by means of a circulating refrigerant through the compressor and to the outdoor coils. In cold weather, a reversing valve changes the direction of refrigerant flow and heat from the outside air is transferred indoors.

The name heat pump describes exactly what the system does. It pumps heat from one place and puts it in another. The cooling part of the system is easy enough for most people to understand. However, it seems strange to many people that heat can be pumped inside when it is below freezing outside. But cold is a relative term, describing a relative lack of heat. Until the temperature hits absolute zero, there *is* heat present in the air. (Just think, it could be even colder outside, so that outdoor air is warm compared to some lower-temperature air.)

The single unit serves all of a home's heating and cooling needs, except during periods of extreme cold (below 0 F) when the efficiency of the system decreases and auxiliary heating coils may be necessary. These are often located within the ducting, and they work exactly as do the heating elements in any other appliance.

Once the thermostat is set for a comfort position, the heat pump either cools or heats as needed to maintain that temperature constantly, year around. Indoor air is circulated through the home by means of one or more blowers) through the duct system.

Many heat pumps have a dual-purpose, automatic thermostat. On days in the spring or fall when the thermostat setting may have to be changed several times, kicking in the heat or switching the cooling unit in, these heat pumps will efficiently maintain a constant air temperature by doing all the switching automatically.

In areas where winter temperatures fall considerably below freezing, most homeowners install electric baseboard heating strips to serve as back-up heating elements when the heat pump is not sufficient to warm the home by itself.

The problems and methods of troubleshooting the heat pump is essentially the same as with a standard air conditioner. Manufacturers claim that unit breakdowns occur less often because the compressor never sits idle for long periods, and is thus less apt to seize up.

If the system uses auxiliary heat strips, troubleshooting, repair, and servicing is the same as with any heating element. It might take you some time to locate the access panel for the heating strips (call the installer), but once you have, the elements can be tested for voltage supply and continuity with your VOM.

ROOM AIR CONDITIONERS

Room air conditioners can be set into a window or fitted into a through-the-wall sleeve. Although smaller, they are the same as any standard house air conditioner in basic construction and operation. Complex units can cool, dehumidify, filter, clean, and circulate the cooled air. Wall and window units vent heated air from the motor/condenser to the outside. Most models also draw outside fresh air in through the coils during the filtering/cooling cycle, eliminating stale air problems.

The basic operational principal of a typical room air conditioner is as follows: warm air from the room is drawn over the cooling coils, giving up its heat. The conditioned air is then recirculated by a fan or blower. Heat from the warm air causes the cold liquid refrigerant which flows through the evaporator to vaporize. The vaporized refrigerant carries the heat to the compressor. This device compresses the vapor and raises the temperature of the refrigerant to a point higher than the outside air. In the condenser, the hot vapor liquefies and gives up the heat from the room air to the outdoor air. The high pressure liquid refrigerant then moves through a restrictor, which reduces its pressure and temperature. Now cold again, the liquid refrigerant re-enters the evaporator, and the entire cycle repeats over and over.

The average room air conditioner has a control to turn the unit on and off, and another to regulate the temperature. Knobs and levers are provided for vents, air deflectors, exhausts, or other special features of the individual unit. These controls are usually on the front panel, or concealed behind a snap-open panel or plate. They may, however, be located on the top or sides of the unit.

With most integral chassis designs, the outer cabinet is fastened permanently to the chassis. Most components can be serviced by partially taking apart the control area without taking the air conditioner out of its mounting location.

Another type has a slide-out chassis that is removed for service work, and the outer cabinet is left in the wall or window.

Common to most units are thermostat controls, multiple-speed fan switches, and condensation disposal to get rid of any water removed from the air. Some models have manual air direction controls; other models have motor controls to the louvers in these vents, automatically oscillating the air movement. Most units have overload controls built in as protection against overheated motor windings. Filters remove any dust in the air. These may be of metal mesh or plastic construction (which can be cleaned and reused), or of glass fiber (which must be replaced on a regular basis).

TROUBLESHOOTING

A common indication of air conditioner malfunction is when room air is warm and musty-smelling, even with the system's compressor on almost continually. The two most common causes are (1) a clogged filter and (2) frozen pipes or fittings in the refrigerant circulating system.

Clean or change the filter, depending upon the type of filter and the manufacturer's instructions for your particular unit. For central air conditioners, the filter is probably in the house in the air return opening. In window units, the filter is usually located in a door to the air return and next to the condenser coils.

If the filter has become clogged with dust and grime, air circulation can be impeded to the point where the unit can no longer operate efficiently. The most common filters are easily replaceable for a cost of only a dollar or so. Examine the filter regularly, and replace it as soon as it seems to be getting dirty. It's also a good idea to use a vacuum or an air blower to clean the surfaces of the air return before reinstalling a clean filter.

Carefully examine the radiant fins. If they are clogged or excessively dirty, clean them thoroughly and carefully. Cleaning the fins by blowing air across them is the best and safest method. With the power off, you can also clean dust and dirt away from the inside of the unit.

Another common cause of poor cooling is a low refrigerant level. A quick, although inaccurate, check is to look at the sight bubble while the unit is operating. You should see a flow of refrigerant passing by the glass.

Check the operator's manual to find out how to test for low coolant levels and to see if coolant replacement is possible from one of the refrigerant spray cans available at most air conditioning or hardware stores. Often coolant replacement requires special gauges and equipment, however. You might prefer to pay a technician to come out on a regular basis to check and refill the refrigerant. At the same time (and usually for the same cost) the entire unit will be inspected and cleaned.

If you choose to handle your own servicing, before refilling with refrigerant, follow the manufacturer's instructions on how to test for leaks in the refrigerant lines. If leaks are found, they may sometimes be repaired by use of one of the chemical resealing sprays (check with your dealer), or by completely draining the system and silver soldering or brazing the leaking line. This should not be attempted unless on the direct recommendation of the manufacturer or a qualified service expert. The line carries high pressure and can be dangerous.

Don't ignore a low refrigerant level. If too low, it can burn out or oth-

FIG. 12-8 Check the sight glass to see if the refrigerant level is low.

erwise damage the compressor in a very short time. The cost of a service call is minor compared to that of replacing a compressor.

Even when it is 110 F outside, the air conditioner might freeze. In this case the thermostat will have to be turned off until the refrigerant tubing thaws. Occasional freezing is nothing to worry about. If the freezing is a daily or a frequent difficulty, several possible factors must be considered. One is the filters. A clogged filter leads to system inefficiency and potential freezing. Clean the filters thoroughly, or replace them. If you are cleaning plastic or metal mesh filters, be sure they dry thoroughly before you replace them and turn the air conditioner back on. Do not allow the unit to operate while you are changing or cleaning filters — this is an invitation to dust to enter the aluminum radiator fins.

Some technicians recommend that the blower fan, which normally kicks in only when the compressor goes on, be switched to manual and left on all the time — whether the compressor is on or off. Other experts say this should never be done, since the constant running of the blower fan can *cause* a freezing problem. Others say to leave the fan on automatic but shut the system down completely for a short period of time (15 minutes to a full hour), once or twice a day.

There is no one solution to freezing that works for all systems. The consensus appears to be that each situation is unique. The cure for intermittent freezing depends upon humidity, the thermostat setting, the size of the cooling unit in comparison to the area to be cooled, the physical location of the compressor/condenser (whether on the roof or otherwise exposed to the sun, or on the side of the house or under some type of shading cover), and upon how extreme the daily temperature swings are in your area.

Your local utility company can be consulted as to the best freeze-up cure for your particular system and locale. Usually such advice will come at no charge.

Although motor problems are relatively easy to diagnose, they are often difficult and expensive to resolve. Most central systems have a compressor motor, a blower or fan motor, and occasionally some type of pumping system for moving the coolant (refrigerant) through the air conditioning cycle (although the pumping action is often handled by the compressor itself). Even room air conditioners can have two or three internal motors. Once you have traced a malfunction down to a particular motor in the system, then determine the operating voltage for that motor and remove it from the unit. Knowing the working voltage in advance can help you make workbench checks on the motor after any repair attempts are made.

In cases where the air conditioner does not turn on at all, systematically check the system's fuse or breaker system and input power wiring. The power train, any voltage switches, and the thermostat should be tested before you assume that a motor is bad. After all, these devices offer easier access for testing in comparison to the heavier and tougher motors.

If all parts of the system test okay except a single motor, then you have traced down the malfunctioning component and you can set about your repair or replacement chores.

The motor trouble can be burnt open or shorted windings, worn or defective brushes, dry (unlubricated) bearings or other moving parts, jammed or bent drive shaft, slipping pulleys or belts (also broken belts, gears, pulleys), or excessive vibration due to loose motor mounts.

Before removing a defective motor, try to turn its shaft by hand (with all power off to the central system). See if there is any mechanical binding, if there are any loose mounting bolts or screws, or if a belt or drive gear is worn or broken. For disassembly and troubleshooting, be sure to refer frequently to the owner's manual.

FIG. 12-9 The compressor motor will probably have both a starting and a running capacitor.

If the motor fails to turn even with supply voltage on, try lubricating the shaft and any accessible bearings. This holds true even if the motor is a sealed, "lifetime self-lubricating" motor. These can often be revived by a judicious application of WD-40 or some other penetrating lubricant. Allow enough time for the oil to soak in thoroughly, and gently work the motor shaft back and forth manually trying to "break" the frozen bearing or shaft that may have overheated due to a lack of oil.

Disconnect power and the voltage input leads of the motor, and use your VOM to test each motor winding for continuity (see Chapter 5). If an open is indicated, be sure to inspect and test the voltage connecting leads to see if a bad solder joint or a broken wire right at the voltage input terminal is keeping the motor from getting power. There are times when a motor with an open winding may be disassembled and the wiring carefully unwrapped turn by turn in the bad circuit. If the open (break) is near the terminal, often a spliced repair may be effected with a soldering gun. As a last resort, try taking the short end of the broken wire and throwing it away and hooking up the end of the long part of the winding back to the terminal. If only a turn or two of wire has been lost in a multi-layer coil, the motor (or starting winding) may still work after this type of emergency repair. However, it's a good idea to replace the entire motor as soon as possible.

Remember that all wiring wrapped around in a coil is insulated, even if the insulation is only a layer of lacquer. This insulation must be carefully scraped away with a sharp knife or razor blade before any soldering attempt can be successfully made. To solder, first pre-tin the end of wire to be soldered, along with the point where it is to be connected. (To pre-tin is to heat the wire or solder lug to a point where applied solder will melt, then let the solder flux into a silvery coating of the tip or joint.) Then, when the end is twisted around the lug or another wire end for splicing, soldering the connection is easier and faster and the resulting splice is both physically and electrically stronger.

Sometimes a motor that is binding from overheating or poor lubrication can be brought to life by thoroughly cleaning it, lubricating it well (possibly with one of the new silicon based lubricants), and putting it back together again.

When repairs are impossible, look to your operator's manual for a description and part number for the defective component. Order the replacement directly from the maker or from one of the authorized parts/service outlets in your community.

Even if the repair is major, such as a blown compressor, it can save you almost 50 to 75 percent if you purchase a replacement from the manufacturer and handle the labor and installation yourself. This is because a service dealer will mark up the price of the part when he sells it to you, and will charge you highly for his expert technical ability.

At the same time, proper diagnosis and repairs will often require special equipment. It is not wise to attempt such a money-saving service job yourself until you are absolutely sure you know what components are bad, and you are sure that you can properly install the replacement (or can obtain expert advice in your fix-up effort).

Motors, fans, belts, thermostats, switches, air direction controls and vanes, pumps, and many leaking fittings can be adequately fixed yourself. Without prior training and experience, it is best to leave compressor, condensor, evaporator, and restrictor problems to a professional.

Never neglect to check the duct and register portion of your system. Dust, rust, and oil spills can create trouble. Make sure all vents operate properly and that air flow moves freely into each room you want it to and not into rooms you do not wish to cool.

VOLTAGE SUPPLY

It is essential for efficient operation that the electrical service provided to the air conditioner be maintained within 10 percent of the rated outlet voltage. Check the voltage of the power source when the compressor is running. If the utility company in your area has difficulty in maintaining this stability, it is advisable to install a voltage regulator between the electrical service and the air conditioner.

Another safety feature worth installing is an automatic trip device that will completely turn off the air conditioner in case of brown outs, (temporary short-term power failures). Often, power from a utility company may drop to very low levels during periods of peak demand or during power outages when voltage feeds to various neighborhoods are automatically rerouted. This all means that a voltage supply that is supposed to be 120-ac (or 240-ac) may suddenly drop completely out, and then come back at a level 15 to 30 percent below what is supposed to be supplied, then gradually come up to the proper level.

Also common are voltage spikes and surges. These are sudden and often drastic increases in the incoming voltage level. For a fraction of a second, the lines might be carrying several thousand volts.

These sudden changes in voltage can severely shorten the life of the compressor in an air conditioner. It is often better to have a safety switch-out device that will turn the unit completely off in case of a power failure, or when the voltage falls more than 10 percent below the rated levels. Better to suffer a hot house than to find that your thermostat has repeatedly tried to turn the unit back on with insufficient voltage and that the resultant drag has blown the compressor completely out of service.

EVAPORATIVE COOLERS

Evaporative coolers consist of several pads of excelsior-type material kept damp by a continuous dribble of water. A fan blows or sucks air across these wet pads and then to a ductwork system feeding the rooms of the home.

An evaporative cooler works because evaporating water takes away heat and makes things cool. For this reason, an evaporative cooler or precooler is only useful in dry climates. It relies strictly on evaporation. A small pump takes water from the reservoir and moves it through a central tube and then through spider tubes that lead to troughs over the pads. The water then leaks down and flows over the pads. Excess water dumps back into the reservoir.

The water supply is kept at a fairly constant level through a valve and float that is similar to the one in your toilet. As the water in the reservoir reaches a certain level, the float shuts off the incoming water. As the level drops, the float lowers and opens the valve, allowing more water to come in.

Meanwhile, a larger motor drives a squirrel-cage type of blower, usually with a pulley and belt arrangement. This motor often has two speeds, and sometimes three. It blows air into the home from the box of the cooler. This causes air outside the box to be sucked in through the water-soaked pads, thus cooling the air. Because the amount of air flow is critical to operation, windows must be kept partially open to allow the incoming air to escape without creating excessive back pressure. (If the incoming air can't escape, the air in your home will also be excessively humid.)

There is very little to go wrong with an evaporative cooler. Diagnosis is almost always quick and easy.

If nothing at all is working, power is probably not getting to the cooler. It's unlikely that the main motor and pump will both go out at the same time. Testing for power is usually simple. The cabinet of the cooler is designed to be opened easily, which makes access a simple matter.

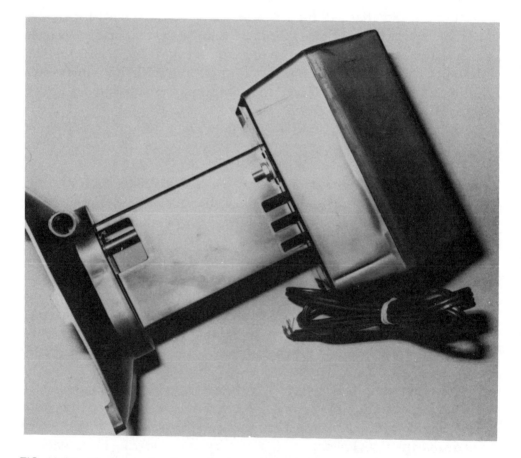

FIG. 12-10 The pump for the cooler is easily and inexpensively replaced.

The pump probably plugs in. Turn the switch to pump only and carefully use your VOM to prove the socket. (The motor might plug in also, but it is more likely to be wired directly to the switch in the home.) You now know that either the fuse or breaker is blown, or the switch is bad—and both are easily fixed.

The only thing that might be slightly difficult is changing a bad motor. The new motor probably will not come with the cord attached, and you might have to scavenge the old motor or buy a replacement cord. If you get an exact replacement (you should), both the wires and screws will be color coded. Instructions will come with the new motor, detailing the exact steps.

Instructions on the pads say that they should be good for three months. Most people find that the cooling efficiency drops and the humidity climbs noticeably after one month, or less if you are using hard water—and even less if you haven't installed a bleeder kit to keep the water in the reservoir fresh and clean.

Change the pads often during the cooling season. The air will be cooler with lower humidity, and will also be fresher and cleaner. Not only is the cost relatively small, this gives you the chance to clean, lubricate, and check out the unit.

If your cooler is pump fed, the standing water in the base of the unit can become a breeding ground for mosquitoes and flies. Put a couple of teaspoons of light-grade machine oil in the cooler water. This will help keep the water from becoming a depository for mosquito larvae.

FIG. 12-11 The motor for the cooler.

When a pump is used and the pads are dry even when the pump turns, then check the pump filter, hose, and the centrifugal vanes that force water upwards. When the pump seems to be operating okay, but the water level in the base of the unit is consistently low, test the inflow valve and valve float to see that it does not shut off too early. This system is very much like the float ball and valve system in the standard toilet fixture, with the float attached to a lever. When water level rises to a certain point, lifting the float, the lever actuates a water turn-off valve.

CHOOSING AN AIR CONDITIONER

There are two types of air conditioner. There is the complete central system for controlling temperature in the entire home, and smaller individual room units which cool only the area into which they are placed. There are limitations and advantages involved in both types, so the choice to be made depends upon individual need.

When shopping air conditioning, three terms are encountered regularly: *BTU*, or British thermal unit, *horsepower*, and *tonnage*.

A BTU is the heat energy needed to raise the temperature of 1 pound of water 1 degree F. The BTU rating of an air conditioner refers to the amount of heat it will remove.

Tonnage is a rating more commonly used with large air conditioners. One ton of cooling is roughly comparable to the cooling effect achieved by melting a ton of ice over a one hour time period. In general terms, for every ton at which an air conditioner is rated, it will provide approximately 12,000 BTUs of cooling per hour.

Horsepower (747 watts) is an old-fashioned rating system, seldom in use today. Horsepower is developed by the compressor motor, but there are many other components of the modern system which affect the cooling efficiency, so the 1-horsepower (compressor motor) room air conditioner may vary in air cooling capability from as low as 6,000 BTUs to as much as 12,000.

Because of the expense of operating air conditioning systems, many homeowners are switching from pure air conditioning systems to "piggy-back" cooling arrangements. This means the installation of both a regular compressor-type air conditioner and an evaporative-type cooler. In hot, dry climates, an evaporative cooler can maintain indoor temperatures at a level of 80 to 85 F or lower, with the only expenses being the electrical energy required to turn a small fan and a pump in recirculating water systems and the water used in the system. Energy costs for running an evaporative cooler can be as much as 50 to 90 percent less than to operate a standard air conditioner. The evaporative coolers loose their efficiency, however, when outdoor temperatures exceed 100 F, or when relative humidity levels go above 15 percent.

A piggy-back system uses an evaporative cooler for most of the warm weather season, and either automatically or manually switches over to regular air conditioning when the humidity or temperature begins to rise above comfort levels.

Often, duct "valves" are installed so that moist air cannot backflow from the evaporative cooler to the air conditioner and so that cold, dry air from the conditioner won't be wasted by flowing back through the ducts to the evaporative cooler.

Since moisture can corrode even the galvanized metal used in duct construction, many installers prefer to see completely separate duct systems when a central system is designed around both an evaporative cooler and an air conditioner. Room register vents for the unit not in use are closed off, while the vents for the operating system are opened to allow either moist or dry cool air into the home.

ELECTRIC FURNACES

A heat pump, or a heat pump with auxiliary heat strips, can keep a house warm. If used, those heat strips are located in the air-handling system, and the same blower fan can be used to carry the heat to all parts of the house. A separate furnace works in much the same way, except that it will have its own blower motor and fan. What you have, essentially, is an electric oven with a fan to move the warm air to where it is needed.

The majority of residential central-heating systems are fueled by natural gas or oil; some, however, are powered by electricity and it is with those that we are concerned here. (Gas- or oil-fired furnaces are best left to someone licensed to handle those systems. The dangers involved are very real — so much so that, in many areas of the country, it is illegal to work on one without the proper certification.) As a rule, the heat is automatically controlled by means of a thermostat, just as most air-conditioning units are. When the temperature goes below the thermostat setting, the bimetallic contact strip energizes an electrical circuit which turns on the heating elements and blowers. Once the temperature reaches the setting on the thermostat dial, the thermostat automatically turns off the source of heat.

There are two basic types of electric heating systems: gravity warm-air and forced warm-air. Gravity systems operate on the principle that hot air rises, since it is lighter than cooler air. A gravity system works efficiently only in homes with basements, or in multi-story buildings, where the central heater is below the normal living level. The heating elements generate heat when electrical energy passes through their coils, and the hot air then rises through ducts. Usually the air movement is augmented by fans. During the upward heat movement, warm air displaces the colder air, which then drifts downward through a series of "cold air ducts" until it reaches the lower part of the central element, where it is heated and again repeats the same cycle.

Gravity warm-air devices are used in floor furnaces and some space heating systems. Several smaller heating elements and associated blowers are placed either under the floor or at the baseboard of the area to be heated. The biggest shortcoming of gravity systems is that there must be a large temperature difference between the upward-moving hot air and the cold air it is to displace in order to create good air movement and equal distribution of heat throughout the home.

Forced warm-air heaters utilize a blower to force air from the heating element through the duct work and registers into individual rooms. Use of this system minimizes the temperature differential between the warm air from the heater and the cold air leaving rooms through return vents. The forced air system allows users to control whole house heating, and to have better control over the heating to individual rooms and areas. Often the building is divided into segments, with each area having its own thermostat and individual blower or auxiliary fan support system.

Forced warm-air heaters require a filter to clean the air as it is being circulated. The filter is generally a pad of loosely packed fiberglass which traps dust and keeps it from being blown back into rooms. Permanent filters of plastic or wire mesh are washable; the throwaway fiberglass types must be replaced. Filters should be cleaned or replaced monthly during the heating season. Do not put it off. A dirty filter will block the air flow, decrease the efficiency of the system, and can even cause extensive damage.

The only visible parts of most home heating units are the registers (or radiators) which allow the heat to flow out into a room. Registers may be used with either gravity or forced warm-air systems. In the gravity type arrangements, the register outlets are usually located low — in the floor or along the baseboard of a room. Gravity system outlets are usually placed along the interior walls of a home, to give better and more economical heating. Floor grills located along outside walls are generally return vents for the cool air being displaced.

The junction between the face of the register (its flange) and the wall should be airtight. Better quality registers come with a rubber or felt gasket, which improves the fit. With forced warm-air circulation, the best type of outlet is the diffusion-type register. Clean grilles, registers, and blower fan blades at least twice during each heating season. If the blades become loaded with lint, air moving efficiency deteriorates. Check and adjust the tightness of the belts from the motor to the fan blades each year, and replace if worn or beginning to slip.

Either round or rectangular conduits are used in the duct work. The construction is generally galvanized metal or aluminum. Air leaks in the system may be reduced by using duct tape, available at hardware stores, to wrap around rusted spots or joints which are not airtight. A careful annual inspection of the ducts will enable you to seal air leaks and increase heating efficiency.

Servicing and maintenance of an electric heating system is simple. Some systems have emergency switches, as well as a standard service entrance fuse or breaker. The emergency switch for an electric furnace is a wall switch close to the heater, and it is generally red. When the heater fails to operate, first check that no one has accidentally turned this switch off. If all switches, fuses and breakers are working, try moving the thermostat up a few degrees. If the thermostat has a day and night setting with a timer, take off the thermostat cover and check the dial to see if a power failure or some accident may have caused the timer to get out of cycle. While this cover is off, check the thermostat contact points. Dirt or corrosion on the bimetallic contact points could keep the heater from being energized. To clean these contacts, pass a business card or a crisp, fresh dollar bill between them. Do not use an abrasive contact cleaner, since it can scratch or otherwise damage the bimetallic points. If the thermostat is a mercury vial-type, no cleaning is necessary. The mercury unit is sealed during manufacture.

If heat comes out of the registers but the blowers do not operate, the trouble is in the fan or fan motor, or there is a leak or blockage in the duct work. In some combination cooling and heating systems without a heat pump, there is a plate which may be slid in or out of the duct nearest the cooler. It is removed to allow cool air to flow through the ducts during summer. In the winter, the plate is inserted in order to keep cool air from leaking into the duct system as it is being used by the heating blowers. Check your system to see if it incorporates one of these plates and, if so, that the plate is in the right position for the heater/cooler in use.

Testing the heating element(s) is accomplished in the same manner as testing any other heating element. If a heating element is found to be defective, purchase an exact replacement from the manufacturer or the local dealer outlet. Installation of the replacement through an access panel is relatively easy.

Troubleshooting Guide

Possible Causes	Solution
Problem: Nothing happens.	
No power	Check fuses or breakers and cables
Thermostat bad	Test; replace
Switch or contact bad	Test; replace
Bad start or run capacitor	Test; replace
Problem: Little or no cooling/heating.	
Filters dirty	Clean or replace
Cooling fins dirty	Clean
Belt broken or slipping	Clean; repair or replace
Insufficient voltage	Test; call power company
Freezing on evaporator	Check; allow to thaw; attempt to find and cure cause
Thermostat bad	Test and replace; test control transformer and circuit, and repair or replace
Bad switch or relay	Test; replace
Refrigerant low	Check; refill; call technician
Aging or bad compressor	Check; call in technician
Aging or bad blower fan	Check; repair or replace

Chapter 13
Hot Water Heaters

Of all major home appliances, the hot water heater is least likely to go bad. It is the most reliable of almost all of the electrically operated devices in your home.

Unfortunately, when an electric hot water heater *does* begin to malfunction, there is little in the way of repair service that you will be able to do yourself. Much of the time, attempts at repair cause more problems than cures.

A bad power cord or plug can usually be repaired or replaced. Defective thermostats may usually be replaced easily enough. However, repair of leaks will often merely make a bigger leak, or might fix the leak temporarily, but even then will probably damage the insulation. A defective heater element can sometimes be replaced, but due to the way the elements are installed in the water heater, that replacement could cause a variety of problems, the least of which is a nagging leak.

HOW IT WORKS

A hot water heater is somewhat like a large thermos bottle. Years ago, the tanks were metal, which rusted sooner or later. Modern water heaters are glass lined, which means they have an internal coating, usually of porcelain. This is an inexpensive way to protect the tank and makes the unit last longer.

To help hold the heat, the tank is surrounded by fiberglass insulation. To protect all of that, and to make the unit more attractive, a metal outer casing is used, with door and access panels in appropriate spots.

With many units, water comes in and goes out through two pipes in the top. These are marked. When installing a new unit, be sure that you

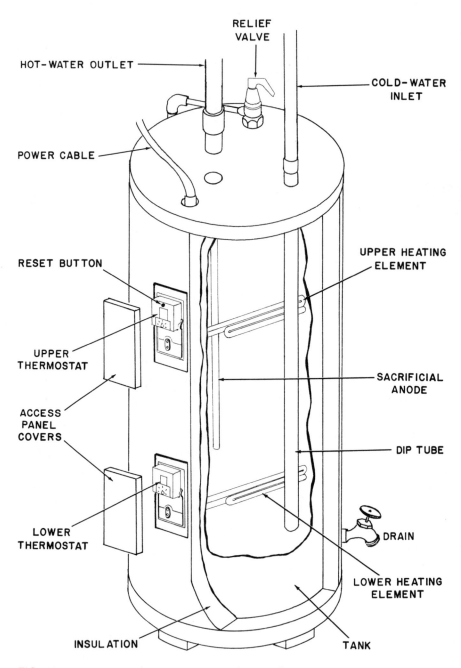

FIG. 13-1 Location of parts in a typical water heater.

Labels in figure:

HOT-WATER OUTLET

RELIEF VALVE

COLD-WATER INLET

POWER CABLE

RESET BUTTON

UPPER HEATING ELEMENT

UPPER THERMOSTAT

SACRIFICIAL ANODE

ACCESS PANEL COVERS

DIP TUBE

LOWER THERMOSTAT

DRAIN

INSULATION

LOWER HEATING ELEMENT

TANK

FIG. 13-2 The top of the water heater.

have the water connected properly. On units with the cold water entering through the top, an internal pipe called a dip tube (often made of breakable plastic) is used to bring the cold water to the bottom of the tank. A few units have the cold water inlet on the side and towards the bottom. Regardless of how the cold water enters, the hot water always comes out the top.

Also in the top of the unit is a temperature-pressure relief valve, mounted where the water is the hottest. Normally this valve just sits there and does nothing. If the internal pressure goes too high, this valve "blows" and releases the pressure before it damages the tank. If the internal pressure builds up sufficiently, the water can go off like a bomb, destroying not just the water heater but also the surrounding parts of your

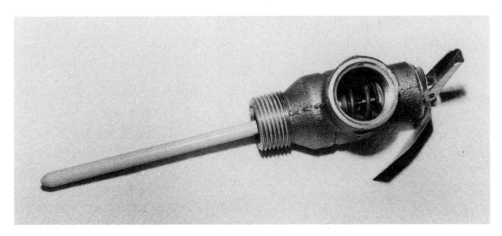

FIG. 13-3 The water heater relief valve.

home. Sometimes this valve will be connected to a pipe that conducts the steam and water to a safe outlet. Preferably this relief pipe should not go outside the house, especially in cold climates where it might freeze.

At the bottom of the tank is the drain valve, which allows you to empty the tank for repairs or cleaning. (Don't forget to shut off the incoming water before you open this valve.) The bottom of the tank is often concave, which gives strength and makes flushing through the faucet more efficient.

Also towards the bottom is the access panel for the thermostat. One or more screws hold this in place. After removing the holding screw(s), the panel has to be moved down to allow it to slip out.

Inside are the heating elements. There are usually two, and sometimes three. In a few units, the heating elements are removable and may be replaced if faulty. Remember, though, that most modern hot water heaters have two or three heating elements, and each will have to be checked in case of a failure in the system.

The heating elements are often held in place by large bolts in the side of the tank, near the bottom, middle, and top of the heater. The holding bolts are mounted through sealing rubber washers, and these should be replaced and plumber's caulk used for added moisture-proof reinstallation.

The cold water enters and flows through a dip tube that takes it to the bottom of the tank (unless the unit has the cold water inlet at the bottom, of course). Quite often, at the top of the dip tube is a smaller opening, designed to prevent water from leaking back into the household supply. At the bottom of the tank, the water is heated by the lower heat-

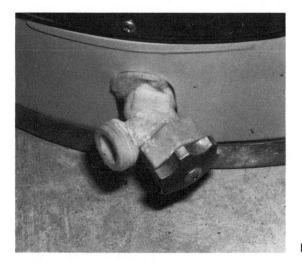

FIG. 13-4 The drain valve.

FIG. 13-5 Removing the access panel.

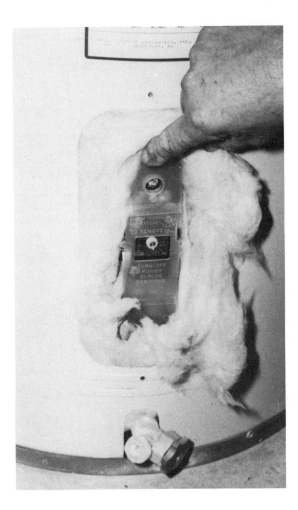

FIG. 13-6 Water heater thermostat device. Note reset button.

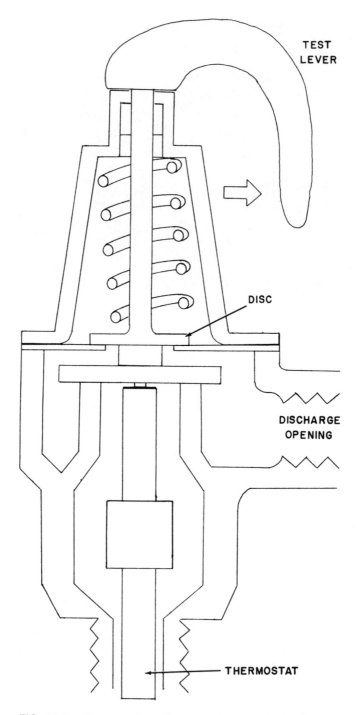

TEST
LEVER

DISC

DISCHARGE
OPENING

THERMOSTAT

FIG. 13-7 Cross section of a water heater relief valve.

ing element. As it is heated, the water moves up to the top of the tank where it is kept hot by the upper heater element.

Power enters the water heater through a reset device – the high-limit protector. If the temperature exceeds a certain preset limit, an automatic switch is flipped, cutting off all power to the water heater. A reset button is used to put the hot water heater back into operation.

Power then enters the thermostat for the upper element. This is set up like a double-pole single-throw switch. When power is needed by the upper element, all power to the lower element is cut off. Once the water in the upper part of the tank is hot enough, the upper element is cut off and power can then be supplied to the lower element where initial heating takes place.

For the lower element, a second thermostat comes into play. If the temperature in the lower section of the tank is below the set level, this element activates and warms the water. The warmer water then rises in the tank.

DIP TUBE PROBLEMS

As mentioned, in many units the cold water enters through the top and is brought to the bottom of the tank through an internal pipe called a dip tube. This tube can corrode through or become loose, resulting in cold water being fed to the top of the tank instead of to the bottom. The problem is called bypassing, and usually means all you get from the house faucets is a supply of warm water.

If the water from the faucets in your home is only warm at best, or if you run out of hot water too quickly, spend a few moments to examine the water heater. You can often detect dip tube problems with nothing more than your hand.

As hot water is taken from the tank, new cold water is fed into the tank at the bottom. Preferably with power removed – to make sure that the lower heating element doesn't warm the water – draw off some of the hot water (about a half-tank full), open the access panel and feel the bottom of the tank. It should feel cold, while the upper part of the tank should feel hot. The line between the two parts (cold and hot) should be sharply defined.

The cure is to replace the dip tube, or to replace the entire water heater. If you opt to replace just the dip tube, be sure to use good plumbing practices to prevent leaks.

FIG. 13-8 To replace the dip tube, you will have to loosen the unions or cut the pipe.

LEAKS

The water in a tank holds many dissolved minerals that can corrode the heating element bolts. Removal of the bolts, even using the utmost care, can lead to tiny punctures of the tank. And these leaks, even if only pinholes, cannot be successfully or safely repaired. Once the tank begins to leak, the heater will have to be replaced. For this reason, it is recommended to replace the entire water heater instead of replacing a heating element. Many water heaters don't even have access panels to get at the heater elements due to the trouble that can be caused by attempts to replace the heater elements.

Some leaks *can* be fixed. User repairable items include the drain valve (faucet), which may become leaky or jammed; the temperature-pressure relief valve at the top of the unit (to vent steam pressure in case of excessive heating of the water); the water inlet or outlet pipes or pipe unions; and sometimes the cold water dip tube.

These are standard home plumbing repairs. Replacement parts are generally available at most hardware or plumbing stores. If an inlet or outlet pipe has developed a single leak in a certain spot, shut off the power to the water heater, preferably let it cool, turn off the input water feed valve, and use a garden hose connected to the drain valve to empty the hot water tank.

Using a hacksaw, carefully cut out the leaking section of pipe, or disconnect the entire length of the faulty section from its union (joint) connections. If a bad section is cut out, a pipe threading tool may be used to cut new threads on the two ends exposed by your cut. A short union (straight or elbow union) may be used to replace the cut out section if it was quite short. If the damaged pipe section was longer, then you will have to purchase a length of pipe long enough to fill the gap and two unions in order to connect the replacement to the existing plumbing.

Remember to use liberal applications of plumber's caulking compound when connecting water joints, and study the repaired junctions carefully for several days after water pressure has been applied to make sure your repairs were not faulty.

Replacement drain faucets and temperature-pressure relief valves are easily installed. Instructions generally come with the part. If you have questions, the store selling the parts can generally provide advice concerning proper replacement steps. Again, be sure to replace all washers with new ones when making any plumbing changes, and apply plumbing compound to the screw-threads before attaching the new valve.

In emergencies, it is sometimes possible to make at least temporary repairs to leaks on a water heater storage tank. There are special silicone and glass adhesives, and a brazing torch may be tried. Remember, though, that there is still a leak on the other side of the tank insulation in the glass lining. The resultant seepage will soon make the entire heater go bad. Worse yet, the continuing seepage can create a dangerous electrical situation.

THERMOSTATS

There are two types of thermostats used in most electric hot water heaters. One type is the rheostat-controlled bimetallic element that is usually attached to a panel at the side or bottom of the unit. An exact replacement will probably be necessary, either from the manufacturer or a service representative in your area.

The second type of thermostat looks sort of like a dull silvery colored lump on a wire. It is also bimetallic, but it is sealed in a soldered case. This type is not variable, and it will operate the internal on/off

FIG. 13-9 A typical water heater thermostat.

switch at a temperature predetermined by the manufacturer. These thermal switches are found more often in gas water heaters than electric. Replacements are readily available at any plumbing supply store at a very reasonable price.

With the power off, you can test the thermostat for continuity (see Chapter 6). Replacement is usually a matter of unplugging and snapping out the old unit and merely reversing those steps for installation.

HEATING ELEMENTS

Getting at the heating elements can be a chore, especially the upper element. This sometime entails removing the outer casing and then cutting through the insulation. And as already mentioned, due to corrosion the replacement of the element often means that the water heater is going to develop leaks.

Although replacement is possible, it is generally recommended that if the heater elements are gone, it's time to replace the water heater as a unit. By the time the elements burn out, chances are pretty good that the water heater is getting old and will have to be replaced soon anyway.

The lower element is generally located by the thermostat assembly. The upper one is often directly above this and about one fourth of the

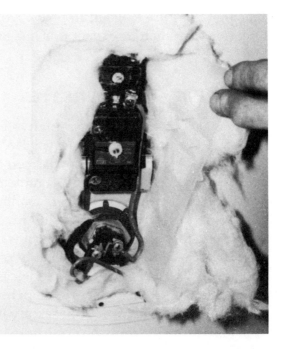

FIG. 13-10 The wires enter the lower heater element.

way down from the top of the tank. If an access panel isn't supplied, you will have to remove the outer case. *Note:* Many water heaters do not have an upper element.

The heater element is usually bolted to a watertight flange which in turn is permanently welded to the tank. Double check to be sure that the power is off and that the tank is drained. Then, very carefully remove the

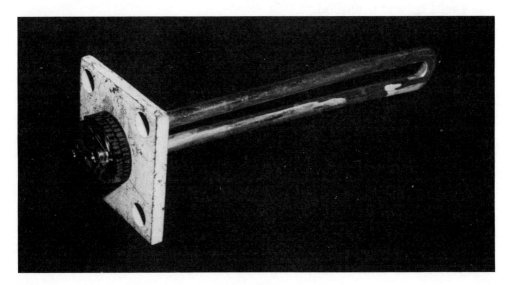

FIG. 13-11 A typical water heater element.

bolts that hold the element in place. Expect some problems with corrosion. (For that matter, expect to damage the water heater.)

REPLACING A WATER HEATER

When a water heater fails, it is relatively easy to install a replacement of the same or even larger size, especially with modern flex-hose feed systems. If the original appliance that you are removing is of the old style, fed by a rigid piping system, complete adaptor kits using the water input/output flex-hose system are available at plumbing outlets for less than $20. You may find conversion to this newer tubing essential even when replacing the old unit with a new one of the same capacity. Newer water heaters are designed physically smaller, especially with regard to unit height, as opposed to older models. A new 50-gallon heater could be as much as 6 to 12 inches shorter than the one you intend to replace, and considerable refitting will be necessary in order to connect that new unit to the old fixed plumbing.

Before removing a damaged unit, turn off the input water supply and throw the service entrance circuit breaker feeding electricity to the water heater. Attach a hose to the drain valve and empty the tank.

Now you can take apart the unions in both the cold input and hot output pipes. If there are no unions, with the pipes going directly from the heater into a wall, then use a hacksaw to cut the pipe, threading the stubs and installing unions (or new flex-hose couplers). Have plenty of towels around. Regardless of how well you drain the lines and tank, there will be water spilled.

Turn the emptied and disconnected tank on its side and roll it away. Be sure to scavenge the "retired" unit for still usable relief, pressure, and drain valves, or of any other electrical or plumbing fixtures which may be of use in future repairs of the new water heater. Other than these spare parts, the old unit probably has no value.

If the new unit is of different height, make the necessary additions to the water pipes to connect up to the input and output unions. Reconnect all electrical wires, according to the manufacturer's installation instructions. *Do not* turn the power back on again until the tank has filled completely!

Turn on the input water supply and then turn on a hot water tap somewhere in the home and let it run for awhile in order to refill the water tank. Then turn the electric circuit breakers back on, and set the thermostat controls to the desired temperature.

TEMPERATURE SETTING

Today, with energy costs going up and up, people are justifiably concerned with the temperature setting on their water heaters—the single largest draw in most homes. Setting the temperature too high can cause problems beyond just high energy costs. For one, excessive temperature can cause excessive pressure— approximately 10 pounds of pressure for each degree rise in temperature. The life of the water heater is then lessened, while the danger is increased.

On the other hand, setting the temperature too low can also cause problems. The first place you might notice this is in running out of hot water while taking a bath or shower. Since the temperature of the water is lower, less cold water is used to mix with the hot—and the result is that more water is being drawn from the water heater. The more that is taken out, the less there is.

FIG. 13-12 Setting the temperature on a water heater.

Another problem with a low temperature setting is that appliances such as dishwashers require water of a certain temperature to operate well. If the temperature of the water used in the dishwasher drops below about 140 F, the dishes won't come out as clean.

It might take a bit of experimenting before you finally settle on a temperature setting. A good place to start is right in the middle, with a setting of around 150 to 155 F. If it seems plenty hot for you, try reducing it slightly (say to 145 F). That will probably be too low, but it's worth a try, considering the savings that will result.

MAINTENANCE

There is very little to maintain on a water heater. It is one of those things that either works, or it doesn't. And when it fails, the cure is generally to replace the entire unit.

However, you can increase the life of your water heater, and increase its efficiency, by a few simple things.

The best way to increase efficiency is to shop wisely in the first place. All water heaters these days have efficiency rating tags, and the various makes and models are easily compared.

Beyond this, the best way to increase efficiency is to add insulation to the outside of the tank and to the pipes that carry the hot water around your home. The more exposed the water heater and pipes are, the more important it is to insulate them.

Once or twice per year the tank should be cleaned. In the past, when all- metal tanks were used, this was a critical step if you wanted the tank to last more than a few years. With modern glass-lined tanks, cleaning out the tank isn't quite as important, which is probably why so many people today forget to do it. Glass-lined or not, the tank will have a slow buildup of sediments. And there are still metal parts inside that can corrode. A regular flushing will increase the life of the tank and its efficiency.

To do the cleaning, first shut off the power to t he tank. For safety, let the tank sit and cool for a while. If the drain at the bottom isn't connected to a pipe (they rarely are), you can connect a hose. Be sure that the hose has a gasket in good condition and that the coupling is tight enough to prevent leaks. The tank will drain by gravity, so be sure that the other end of the hose is lower than the tank and is outside and away from anything that could be damaged by water. Now all you have to do is to open the drain valve.

Some sources say that you should completely shut off the incoming water before beginning. Others say to leave it on for a short time to help flush the sediments. Still others say that the incoming water should be shut off at first, and then turned on again after the tank has completely drained and while the drain valve is still open. A few sources even suggest that you shut off the incoming water, drain the tank, shut off the drain valve and open the incoming water to refill the tank, and shut off the incoming water and drain the tank again.

Sediments and normal corrosion can also be a problem concerning the temperature-pressure relief valve. This valve protects both the water heater and your home. If pressure builds up too high in the water heater, it can literally explode. The valve prevents this. Many sources suggest flushing this valve once or twice each year. Most have a handle to allow this. As common sense as this might sound, many people find out that the relief valve is clogged *after* the tank has burst.

Exercise caution when flushing the relief valve. Since it is usually done while the water inside is hot, it is all too easy to get a nasty burn. If the valve isn't already connected to a pipe, you can use a hose, just as with draining and cleaning the tank. Exercise caution either way.

Inside some water heaters is a magnesium anode. The purpose of this tube of metal is to help prevent damage from electrolysis (interaction of metals). Magnesium is more reactive than the metals of the tank. Consequently, instead of the force of electrolysis eating away the tank, it eats away at the anode, with the magnesium then being coated on the sides of the tank. Occasionally this anode should be inspected. If need be, replace it. To do either, shut off all power and drain the tank. The old anode will unbolt, or unscrew, from the top.

FIG. 13-13 A regular testing and flushing of the relief valve is a good safety precaution.

Troubleshooting Guide

Possible Causes	Solution
Problem: No hot water.	
No power	Check breaker, incoming wires, and all contacts and connectors
Thermostat (usually upper)	Test; replace
High-limit protector	Test; replace
Heater element(s) bad	Replace water heater
Problem: Insufficient heating; not enough hot water.	
Tank is too small for load	Reduce load, or get larger tank
Thermostat (usually lower)	Test; replace
Heater elements bad	Replace water heater
Sediments in tank	Flush tank
Dip tube bypass	Check dip tube
Problem: Leaks.	
Holes in tank	Replace
Holes around fittings	Reseal, replace
Excessive pressure or temperature	Check relief valve, replace; reduce temperature
Problem: Noises.	
Sediments in tank	Flush tank
Water heater improperly installed	Check installation, reinstall

Glossary

ac Alternating current. Supplied by power companies to homes. It is electricity moving back and forth along a wiring circuit, first in one direction and then in another.

amp Ampere. The unit for measuring how much electricity flows past a given point in one second of time when pushed by one volt of electrical pressure. It is somewhat similar to gallons per minute when measuring water flow through a pipe.

analog A signal that varies directly in proportion to another signal. An analog meter uses a needle that swings anywhere across a scale; compared to a digital meter which gives a reading in discrete numbers.

aquastat A device used to measure and control the temperature of the water in a hot water heater, or in a furnace that uses hot water.

armature The part of an electric motor that turns. The *armature winding* carries the current. Also the moving part of a relay.

armored cable More commonly called *BX*. Electrical wires encased in a flexible metallic sheath. Correctly called Type AC.

backflow Water moving in a direction other than normal, or from an external source other than the usual one.

battery A device that holds and/or generates dc power via a chemical reaction.

bearing A supporting part that holds a motor shaft and allows it to spin.

bibb The correct name for a faucet with threads. For example, an outdoor faucet that accepts a hose is correctly called a hose bibb.

bimetal Two metals binded together in a strip, with different bending temperatures. The two act as a temperature sensitive switch to control heating or cooling appliances.

breaker A type of "fuse" that does not melt, but that opens an electrical switch when an overload trips it. Somewhat similar to a switch. Can be reset.

brush A conductive "plug" that makes contact between the commutator of a motor and the source of power. The brush is usually made of graphite or carbon.

bushing An insulator to be placed around the wires extending from the end of a BX cable, to prevent the cut edges of the metal sheathing from damaging the insulation of the wires extending beyond.

BX More correctly called Type AC. A metal-armored cable for use in areas that are always dry. Often used in areas where Romex does not meet building codes.

cam An irregularly shaped device that when rotated on a shaft causes a switch or other mechanism to be tripped.

capacitor An electronic component that stores a charge. It also passes ac while blocking dc. Commonly used as a booster for motors and in power supplies.

centrifugal switch A switch that works as the speed of a motor reaches a certain level, usually to cut out the startup winding and to cut in the main winding.

check valve A valve that prevents water from flowing in the wrong direction. Sometimes incorrectly used to describe a safety valve that "lets go" when too much pressure builds up, such as in a water heater.

circuit A complete path for the flow of electricity, from the source (breaker box or wall outlet), through the appliance, and back to the source through the second connecting wire of the wall plug.

circuit breaker A modern replacement for-old style fuses, this is a switching device designed to automatically turn off whenever a circuit overload or short circuit tries to draw more current than an electrical device is supposed to use. The circuit breaker is simply reset to return electricity to the circuit.

compressor A device used to compress a gas, such as freon. The expanding gas can then be used to cool.

commutator A part of the armature of a motor. Connects to the windings (coils) of the armature. A brush or set of brushes connected to the power supply makes contact with the commutator, and thus with the motor. Most commutators are made of pieces of copper on a central hub of steel.

condenser An old-fashioned term for a capacitor. More commonly, the part of a heating or cooling system that causes a gas to condense to a liquid.

conduit Metal or plastic tubing, used to run cables outdoors or underground, or in areas where exposed wires are in danger of fraying, cutting, or getting wet.

cps Cycles per second.

current The movement of electrical energy in a wire or circuit. Measured in amps.

dc Direct current. The type of electricity provided by batteries and needed for the proper operation of some small appliances. This is current that moves in one direction only, from the source, through the appliance, and back to the source. Some appliances designed to operate ac or dc have a built-in converter, or a power supply, to change ac from the power company into the dc needed to make the device function.

digital Signals with a discrete, on/off state, as opposed to analog signals, which have a continuously changing state.

distribution box Also called a *junction box*. A metal or plastic box, mounted in the wall, through which the wires carrying the incoming power enter the house. These wires are then connected to the wires that run to various outlets, switches, and appliances that operate from that particular circuit.

DPDT Double-pole, double throw switch.

field winding The coil of wire in a motor or other device that generates a magnetic field when current is applied.

fin comb Small pieces of metal in a pattern to disperse the heat generated by an appliance.

fuse A safety device to prevent electrical overloads, and to keep short circuits from causing wires to overheat to the point where they become fire hazards. Fuses are rated according to the amount of current (amps) they can carry, and when a circuit attempts to draw more current than the rated fuse value, it will burn out.

gauge Unit for measuring wire size. The lower the gauge number, the larger the wire, thus the greater current carrying capacity.

GFCI Ground fault circuit interrupter.

governor A device used to control the speed of a motor.

ground Actual earth, or the electrical ground or chassis point of an electrical circuit that provides the equivalent of earth. Sometimes used to refer to the return path of a circuit.

horsepower A measurement of physical work accomplished by an electrical device. When 746 watts of electrical power is used, it is equivalent to one horsepower.

hot wire   The "high" side or power-carrying wire of a home wiring circuit, usually identified by a black or red colored insulation. Neutral or ground wires are usually white.

IC   Integrated Circuit, or IC chip.

insulation   A protective nonconducting coating of rubber, plastic, cotton, or lacquer, used to prevent electricity from jumping across one wire to another.

live circuit   A circuit that is energized, or is carrying electrical current.

load   What is being driven or powered by a device such as a motor. Also used to describe the amount of power used.

mercury switch   A switch that uses a bubble of mercury to make or break contact when the containing tube is tipped.

microswitch   A very small switch, or a switch with a very small contact for activation.

Ohm's law   The mathematical formula relating voltage, current, and resistance. The three basic formulae are: $E = IR$, $I = E/R$, and $R = E/I$, where E is voltage, I is current, and R is resistance.

open circuit   Sometimes called an *open*. A circuit with a break in a wire (like an open hole into which electricity disappears and does not pass through.

overload   The condition when too many appliances or lights are placed on a single electrical circuit, exceeding the maximum amperage capable of being supplied by the circuit. Many household circuits are 15 amps.

polarizing   Identification (usually by color code) of the hot or positive (+) side of a circuit or the ground or negative (−) side of the circuit. These codes help to prevent a repairman from making an accidental short circuit, connecting a hot wire directly to a ground wire.

polarized   A plug or outlet that has a definite ground and hot, or positive and negative electrical or magnetic connections. In polarized plugs and outlets, one prong and one opening will be larger than the other.

power law   The mathematical formula for power, in watts. Expressed as $P = IE$, where P is power in watts, I is current in amps, and E is voltage. If 10 amps of current is flowing with 117 volts, 1170 watts of power are being consumed.

relay   A switch-like device. As current is applied, (e.g., when a built-in timer goes off), the relay causes an internal electromagnet to either make or break contact.

resistance The opposition to the flow of electrical current, measured in ohms.

rheostat An electronic component having a variable resistance. This regulates the amount of current, and thus the power being used. Often used to control motor speed or temperature of a heating element.

Romex A cable covered with plastic or rubber rather than metal.

rpm Revolutions per minute; usually refers to motor speed.

schematic A line drawing showing the electrical connections between various circuits and devices.

service entrance The technical name for the main fuse box or breaker box, where power from the utility company is brought into the house. Total service amperage for most home wiring circuits is 120 to 160 amps, depending upon local housing codes.

short circuit An accidental connection between two points in an electrical circuit, when the hot leg of a circuit accidentally touches the ground leg (or your body) without going through an electric appliance. This provides a path for the maximum amount of current available from the power company, and trips any breakers or fuses. Short circuits can cause sparks and are fire dangers.

solder Generally, a mixture of tin and lead with a relatively low melting point. Used to provide a clean and corrosion-free joint.

solenoid A device in which an electromagnet pushes and pulls at a shaft, causing an inwards/outwards motion, and thus making or breaking mechanical contact. Often used as an electrically controlled valve.

SPST Single pole, single throw switch.

solderless connectors Crimp or twist caps to be placed over exposed wires, making a good electrical connection without solder.

test light A neon or low-wattage bulb with two test-probe leads. Can be used to check for the presence of voltage, ac or dc.

thermocouple A device that converts changes in temperature to electrical energy, often in the form of pulses to control other devices or equipment.

thermostat A device sensitive to temperature changes; it throws an internal switch when certain temperature levels are reached. Most are bimetallic, containing two metals that expand and contract at different rates and make or break electrical contact, depending on how far they are "bent" by temperature variance.

torque Measure of rotating froce around an axis.

transformer A device used to change the value of the ac voltage going through it. The most common purpose in the home is to reduce line voltage to the needed value.

voltage   The electrical "pressure" that moves electrical energy through a circuit. Measured in volts.

VOM   A volt-ohm-milliammeter, for measuring voltage, current, and resistance in electrical circuits.

watt   The unit of use of electrical power. One watt used for one hour is a watt-hour. 1,000 watt hours is a kilowatt hour, the measurement a utility company uses in order to bill you for electricity each month.

winding   The conductive wire of a motor that is wrapped and coupled inductively to the magnetic core.

Index

See also the Glossary, pages 243-248.